3D Integration in VLSI Circuits

Devices, Circuits, and Systems

Series Editor
Krzysztof Iniewski

For more information about this series, please visit:
https://www.crcpress.com/Devices-Circuits-and-Systems/
book-series/CRCDEVCIRSYS

3D Integration in VLSI Circuits

Implementation Technologies and Applications

Edited by
Katsuyuki Sakuma

Managing Editor
Krzysztof Iniewski

CRC Press
Taylor & Francis Group
Boca Raton London New York

CRC Press is an imprint of the
Taylor & Francis Group, an **informa** business

CRC Press
Taylor & Francis Group
6000 Broken Sound Parkway NW, Suite 300
Boca Raton, FL 33487-2742

First issued in paperback 2021

ISBN-13: 978-1-03-209554-7 (pbk)
ISBN-13: 978-1-138-71039-9 (hbk)

Publisher's Note
The publisher has gone to great lengths to ensure the quality of this reprint but points out that some imperfections in the original copies may be apparent.

Library of Congress Cataloging-in-Publication Data

Names: Sakuma, Katsuyuki, author.
Title: 3D integration in VLSI circuits : implementation technologies and applications / [edited by] Katsuyuki Sakuma.
Description: Boca Raton, FL : CRC Press/Taylor & Francis Group, 2018. | Series: Devices, circuits, & systems | Includes bibliographical references and index.
Identifiers: LCCN 2018010530| ISBN 9781138710399 (hardback : acid-free paper) | ISBN 9781315200699 (ebook)
Subjects: LCSH: Three-dimensional integrated circuits. | Integrated circuits--Very large scale integration.
Classification: LCC TK7874.893 .A16 2018 | DDC 621.39/5--dc23
LC record available at https://lccn.loc.gov/2018010530

Visit the Taylor & Francis Web site at
http://www.taylorandfrancis.com

and the CRC Press Web site at
http://www.crcpress.com

Contents

Preface

More and more products are using three-dimensional (3D) integration technology nowadays. Through-silicon vias (TSVs) have been used in high bandwidth memory (HBM) modules and will become mainstream for other high-end products such as graphics processing units (GPU) and high-performance computing (HPC), with applications in databases, security, computational biology, molecular dynamics, deep learning, and automotive. There is no doubt that 3D integration is gaining a significant attention as a promising means to improve performance as it can provide higher interconnect speeds, greater bandwidth, increased functionality, higher capacity, and lower power dissipation.

Currently, the term 3D integration includes a wide variety of different integration methods, such as 2.5-dimensional (2.5D) interposer-based integration, 3D-integrated circuits (3DICs), 3D systems-in-package (SiP), 3D heterogeneous integration, and monolithic 3D ICs. The goal of this book is to provide readers with an understanding of the latest challenges and issues in 3D integration. TSVs are not the only technology element needed for 3D integration. There are numerous other key enabling technologies required for 3D integration and the speed of the development in this emerging field is very rapid. To provide readers with state-of-the-art information on 3D integration research and technology developments, each chapter has been contributed by some of the world's leading scientists and experts from academia, research institutes, and industry from around the globe.

Chapter 1 by Prof. Franzon from North Carolina State University (NCSU) provides a brief review of 2.5D and 3D technology options, including interposers and TSV-stacking technologies. As illustrated by successful commercial 3D products and experimental results of complementary metal–oxide–semiconductor (CMOS) stacks with a copper thermocompression-bonded interface, Chapter 1 discusses the reasons why 3D integration is superior in terms of power efficiency, performance enhancement, and cost reduction.

Chapter 2 by Dr. Li from Cisco Systems describes up-to-date manufacturing technologies for 2.5D and 3D SiP for integrating application-specific integrated circuits (ASICs) with multiple 3D dynamic random-access memory (DRAM) stacks. In 2.5D integration, both ASIC chip and 3D DRAM stacks are packaged in a planar format on an interposer. This type of integration has emerged as another killer application for 3D integration. This chapter also examines large-size high-density silicon/organic interposers, microbump interconnects, warpage behaviors, packaging assembly processes for ASIC and HBM integration, and board-level reliability tests, and characterization of 3D-integrated packages.

Chapter 3 presents architecture, design, and technology implementations for 3D field-programmable gate array (FPGA) integration and was written by Dr. Ramalingam et al. from Xilinx, one of the major players in this area. Multiple FPGA dies are placed side by side and interconnected on a large silicon interposer. The design simulation methodology, reliability assessment, and future challenges are discussed. On these topics, the authors provide an industry perspective based on volume production of the largest 3D-integrated FPGA as of today, which contains 4.4 million logic cells, 600 thousand microbumps, and 19 billion transistors in a 55 mm package.

Chapter 4 by Prof. Koyanagi et al. from Tohoku University, Japan, covers various unique 3D system-on-chips technologies, such as 3D integration using self-assembly and electrostatic bonding, TSV formation based on directed self-assembly with nanocomposites, and hybrid bonding technology using Cu nano-pillar. Technologies for a 3D-integrated CMOS image sensor module for a driver assistance system are presented, and future 3D integration challenges are also discussed.

Chapters 5 and 6 focus on fabrication approaches for 300 mm wafer-level 3D integration without microbumps. For high-volume manufacturing, TSVs are formed after a wafer-bonding process. The leakage current and electrical resistance of TSVs, I_{on}/I_{off} characteristics for field-effect transistor (FET) devices, and the characteristics of wafer-bonding technologies are also discussed. Chapter 5 by Prof. Ohba from the Tokyo Institute of Technology, Japan, deals with the 3D integrations of permanently adhesive-bonded ultra-thin wafers. In Chapter 6, Dr. Skordas et al. from IBM discussed the 3D integration technology based on low-temperature oxide-bonding for integrating high-performance POWER7™ 45 nm silicon-on-insulator (SOI)-embedded DRAM.

Chapter 7 by Ms. Cheramy et al. from CEA-Leti, France, provides the principles, process integration, and detailed overviews of both monolithic 3D ICs (CoolCube™) and Cu/SiO_2 hybrid bonding technologies. Monolithic 3D ICs enable the stacking of multiple transistor layers in the third dimension, with a vertical interconnect pitch in the range of a few tens of nanometers, and the bottom layer can be any CMOS type, be it bulk planar FET, FinFET, or fully-depleted silicon-on-insulator (FDSOI). Potentially, it will be a key technology driver for the next generation of 3D integration. The Cu/SiO_2 hybrid bonding technology enables wafer-to-wafer and die-to-wafer connectivity with a vertical interconnect pitch in the range of a few micrometers. This chapter also addresses the issue of thermal dissipation in 3D integration.

Chapter 8 by Prof. Chen et al. from National Chiao Tung University, Taiwan, offers examples of novel platforms and application demonstrations, including terahertz (THz) optical components, piezoresistive pressure sensors, and flexible neural sensing biosensors using 3D integration technologies. This illustrates that 3D integration technology can be used in a wide variety of applications and that it will open a new era of electronics and sensors that cannot be achieved with conventional 2D microelectronics.

I would like to sincerely thank all of the authors for their hard work and commitment. Without their contributions, it would not have been possible to provide an up-to-date review of these innovative technologies and the challenges in 3D integration. It is my hope that this book will provide readers with a timely and comprehensive view of current 3D integration technology.

Katsuyuki Sakuma
Yorktown Heights, New York

Series Editor

 **Krzysztof (Kris) Iniewski** is managing R&D at Redlen Technologies Inc., a startup company in Vancouver, Canada. Redlen's revolutionary production process for advanced semiconductor materials enables a new generation of more accurate, alldigital, radiation-based imaging solutions. Kris is also a founder of ET CMOS Inc. (http:// www.etcmos.com), an organization of high-tech events covering communications, microsystems, optoelectronics, and sensors. In his career, Dr. Iniewski held numerous faculty and management positions at University of Toronto (Toronto, Canada), University of Alberta (Edmonton, Canada), Simon Fraser University (SFU, Burnaby, Canada), and PMC-Sierra Inc (Vancouver, Canada). He has published more than 100 research papers in international journals and conferences. He holds 18 international patents granted in the United States, Canada, France, Germany, and Japan. He is a frequently invited speaker and has consulted for multiple organizations internationally. He has written and edited several books for CRC Press (Taylor & Francis Group), Cambridge University Press, IEEE Press, Wiley, McGraw-Hill, Artech House, and Springer. His personal goal is to contribute to healthy living and sustainability through innovative engineering solutions. In his leisurely time, Kris can be found hiking, sailing, skiing, or biking in beautiful British Columbia. He can be reached at kris.iniewski@gmail.com.

Editor

Katsuyuki Sakuma is a research staff member at the IBM T.J. Watson Research Center. He has over 19 years of experience of researching 3D integration technologies and performing various semiconductor packaging research and development projects. His research interests include 3D integration technologies, bonding technologies, advanced packaging, and bio-medical sensors.

He has published more than 85 peer-reviewed journal papers and conference proceeding papers, including three book chapters in the semiconductor and electronic packaging area. He also holds over 35 issued or pending U.S. and international patents. He has been recognized with the IBM Eleventh Invention Achievement Award in 2017 and an Outstanding Technical Achievement Award (OTAA) in 2015 for his contribution and leadership in the area of 3D integration technology development. He was also given the 2018 Exceptional Technical Achievement Award from the IEEE Electronics Packaging Society, and the 2017 Alumni Achievement Award from his Alma Mater, the School of Engineering at Tohoku University, for his contribution to 3D chip stack technology development in the electronics packaging industry. He was a core-cipient of the IEEE Components, Packaging, and Manufacturing Technology (CPMT) Japan Society Best Presentation Award in 2012, and the IMAPS "Best of Track" Outstanding Paper Award in 2015.

Dr. Sakuma received his B.E. and M. Eng. degrees from Tohoku University and the Ph.D. degree from Waseda University, Japan. He is currently serving as an associate editor for IEEE Transactions on CPMT. He served as an associate editor of the Institute of Electronics, Information and Communication Engineers (IEICE, Japan) from 2003 until 2005. He has served as committee member of the IEEE ECTC Interconnections subcommittee since 2012, for the IEEE International Conference on 3D System Integration (IEEE 3DIC) since 2016, and for the IEEE International Reliability Physics Symposium (IEEE IRPS) since 2017. He has been a senior member of IEEE since 2012.

Contributors

Boon Ang
Xilinx, Inc.
San Jose, California

Jonathan Chang
Xilinx, Inc.
San Jose, California

Kuan-Neng Chen
Department of Electronics
　Engineering
National Chiao Tung University
Hsinchu City, Taiwan

S. Cheramy
CEA-Leti
Grenoble, France

L. Di Cioccio
CEA-Leti
Grenoble, France

C. Fenouillet-Beranger
CEA-Leti
Grenoble, France

P. Franzon
North Carolina State University
Raleigh, North Carolina

T. Fukushima
Department of Mechanical Systems
　Engineering
Tohoku University
Sendai, Japan

Yu-Chen Hu
Department of Electronics
　Engineering
National Chiao Tung University
Hsinchu City, Taiwan

A. Jouve
CEA-Leti
Grenoble, France

Myongseob Kim
Xilinx, Inc.
San Jose, California

Chandrasekharan Kothandaraman
IBM T.J. Watson Research Center
Yorktown Heights, New York

M. Koyanagi
New Industry Creation Hatchery
　Center (NICHe)
Tohoku University
Sendai, Japan

Woon-Seong Kwon
Google LLC
Mountain View, California

Tom Lee
Xilinx, Inc.
San Jose, California

Li Li
Cisco Systems, Inc.
San Jose, California

Henley Liu
Xilinx, Inc.
San Jose, California

Cheng-Hsien Lu
Department of Electronics
    Engineering
National Chiao Tung University
Hsinchu City, Taiwan

Liam Madden
Xilinx, Inc.
San Jose, California

Takayuki Ohba
Laboratory for Future
    Interdisciplinary Research of
    Science and Technology
Tokyo Institute of Technology
Tokyo, Japan

Suresh Ramalingam
Xilinx, Inc.
San Jose, California

Katsuyuki Sakuma
IBM T.J. Watson Research Center
Yorktown Heights, New York

Spyridon Skordas
IBM Research Division
Albany, New York

T. Tanaka
Department of Biomedical
    Engineering
Tohoku University
Sendai, Japan

P. Vivet
CEA-Leti
Grenoble, France

Kevin Winstel
IBM Research Division
Albany, New York

Ephrem Wu
Xilinx, Inc.
San Jose, California

Susan Wu
Xilinx, Inc.
San Jose, California

Xin Wu
Xilinx, Inc.
San Jose, California

Ting-Yang Yu
Department of Electronics
    Engineering
National Chiao Tung University
Hsinchu City, Taiwan

1

Three-Dimensional Integration: Technology and Design

P. Franzon

CONTENTS

1.1 Introduction

3D and 2.5D integration technologies permit substantial improvement in form factor, power, performance, functionality, and sometimes even cost. Though not providing a direct replacement for Moore's law, 3D technologies can permit a generation or more of exponential scaling in power per unit of performance and other factors.

This chapter is structured as follows. First there is a review of 3D technologies, followed by a general discussion of commercial 3D success stories and technology drivers. Included in that we review 3D logic projects conducted at North Carolina State University, United States, before closing out on heterogeneous integration.

1.2 Three-Dimensional Integrated Circuit Technology Set

There are a variety of technologies that contribute to the 3D technology set. The purpose of this chapter is to review these technologies and not to provide details of materials and manufacturing options. We will review interposers, 3D stacking technologies and monolithic 3D technologies.

Figure 1.1 shows an *interposer*, often referred as a *multi-chip module* in the past. Interposers get its name from their functionality of being placed between the chip and the package. There are two standard ways to make interposers. One approach is to use multiple thin film layers of spun-on dielectrics and metal patterning. While permitting thick metals, it does not permit high density. High density can be achieved by leveraging a legacy back-end-of-line (BEOL) process from semiconductor chip manufacturing. A BEOL process permits multiple levels of planarized wiring, typically 4–6 layers.

A defining technology in 3D integration is the *through-silicon via* (TSV). The TSV goes through the silicon substrate, connecting the front and back sides of the structure. It is created by etching a near vertical hole, lining it with a dielectric before filling it with metal. The wafer is then thinned to expose the metal. Typical dimensions for interposer structures are 100 μm of thickness for the overall structure and 1–10 μm width and space for the metals. TSVs are typically on at least a 100 μm pitch. The microbumps are typically around a 40 μm pitch, whereas the bumps to the package must be on conventional scales at 150+ μm pitch.

Figure 1.2 provides an illustration for some of the key three-dimensional integrated circuit (3DIC) stacking technologies. This shows a three-chip stack, two of which incorporate TSVs. The top two chips are joined with a face-to-face technology (F2F). Here *face* refers to the top side of the chip—where the transistors and interconnect are. Several methods are used to create high-density F2F connections. Microsolder bumps can be used but today

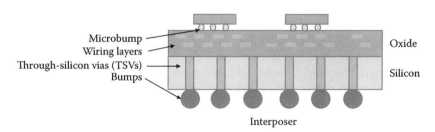

FIGURE 1.1
Interposer (*2.5D*) technology.

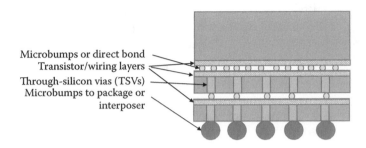

Microbumps or direct bond
Transistor/wiring layers
Through-silicon vias (TSVs)
Microbumps to package or
interposer

FIGURE 1.2
3D stacking technologies.

they are limited to 40 μm pitch (with potential for 25 μm). Copper–copper thermo-compression can be used down to sub-5 μm pitches. Alternatively, hybrid bonding can be used. In hybrid bonding the top surface, with metal plugs in it, is planarized and then bonded to another such surface. An example is the Ziptronix data-based individualization (DBI) process [1]. This pitch can be built down to 1 μm pitch but 6–8 μm are more typical. Hybrid bonding is used to make many cell phone cameras today (Please see Section 1.6 on heterogeneous integration). With such high interconnect densities, many interesting architectures can be explored, as will be explained later.

The bottom two chips in Figure 1.2 are joined using a face-to-back (F2B) arrangement in which the bottom (*back*) of one die is joined to the top (*face*) of another. TSVs are needed to provide the connection to the joining backside. As the sidewalls of the TSV are not entirely vertical, the TSV pitch is limited to approximately the thickness of the wafer, typically 10–25 μm. Thus a F2B connection provides a lot of less density than a F2F connection.

TSVs can be inserted before, during, or after complete wafer processing. These are referred to, respectively, as via early, middle, and last. Figure 1.3 shows a process via middle. Wafers are partially completed, say to metal 1. A vertical side-wall via is etched part way through the wafer. This via is created using the Bosch process [2]. In the Bosch process, deep reactive-ion etching (DRIE) is alternated with a deposition step multiple times to create a near vertical wall via. Today the wall steepness is typically 10:1. Thus the via depth has to be less than 10× the diameter of the opening at the top. The via sidewalls are passivated, typically with an oxide, and then filled with a metal, typically tungsten or copper. The chip metal stack (BEOL) is then completed. The wafers are then flipped and thinned, exposing the TSV metal. The exposed metal can then be used directly for a joining process, or a bump structure is added before joining with another wafer or chip to create a 3D stack.

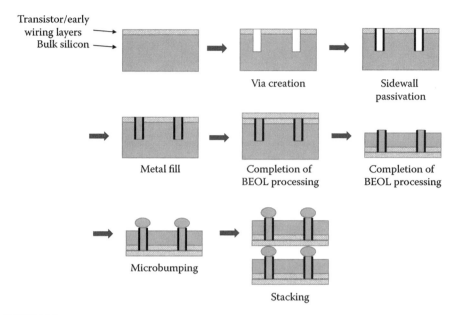

FIGURE 1.3
Through-silicon vias (TSVs) fabrication steps.

Note in Figure 1.2 that the top chip has not been thinned. Usually a 3D chip stack is assembled with the chip in wafer form. Two wafers are attached with the second left unthinned. This wafer stack can be attached to another (thick) wafer for further handling. In most cases, one wafer is left in a thick format so that the wafer stack can be easily handled. Thin wafer handling is possible but increases the cost.

Most 3DICs are assembled in wafer format. Again, the driver is cost. Wafer-to-wafer attachment processing costs less per die and has higher yields. Although chip to wafer attachment is possible it is not widely used.

An intriguing complement to 3D chip stacking technology is monolithic 3D in which there is only one silicon substrate (and thus no TSVs). The most commercially successful one of these technologies is 3D NAND Flash in which the string of NAND transistors in a nonvolatile flash device is fabricated vertically. This approach brings substantial density improvements and cost savings over conventional NAND flash technology. However, to date it has not been commercially applied to any other logic or memory structure.

Another monolithic (-like) 3D technology set involves techniques in which silicon-on-insulator (SOI) wafers are the starting wafer. This is shown in Figure 1.4. In this approach, fabricated wafers are joined face to face using an oxide–oxide bond. As the transistors are built on the top of an oxide layer, a silicon-selective back etch can be used to remove the silicon part of the SOI substrate while not affecting the transistors and interconnect layers. Simple

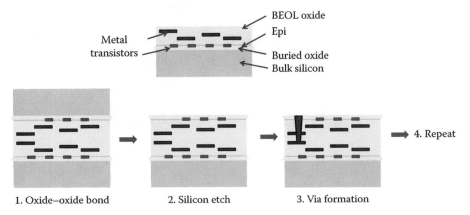

FIGURE 1.4
3DIC chip stack using SOI wafers.

through-oxide vias can then be used to create vertical connections between what were previously separate chips. An example of this process can be found in Reference 3. If the first two chips in the stack are fabricated without interconnect, then one gets two directly connectable transistor layers in what would be considered a monolithic 3D technology. This is monolithic in the sense there is only one bulk wafer left at the end of processing. However, many would not consider it to be truly monolithic as multiple separate wafers have to be created first.

1.3 Three-Dimensional Drivers

In general, the decision to employ 3D technologies is driven by (1) the need for miniaturization, (2) the desire for cost reduction, (3) heterogeneous integration, (4) the need for performance enhancement, and/or (5) the need for improvement in performance/power. Often several of these go hand in hand.

1.4 Miniaturization

An early application of TSVs was providing the I/O connections cell phone camera front-side imaging sensor [4,5]. The goal was not to leverage 3D chip stacks—these were single die—but to reduce the overall sensor height, at least when compared with conventional packaging approaches.

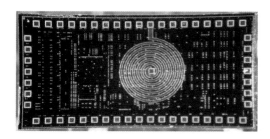

FIGURE 1.5
Miniaturized sensor as a two-chip stack.

3D chip stacking can be used to make such sensors with low integrated volume. Though fabricated using wire bonding, Chen et al. demonstrated an integrated power-harvesting data-collecting sensor with the photovoltaic power-harvesting chip mounted on top of the logic and RF chips [6]. This maximizes the photovoltaic power-harvesting area while minimizing the volume. TSVs and bonding technologies would permit further volume reduction. Lentiro [7] described a two-chip stack aimed at simulating a particle of meat for the purposes of calibrating a new food processing system. One chip is a radio-frequency identification (RFID) power harvester and communications chip, the second is the temperature data logger. It is a two-chip stack with F2F connections and TSV-enabled I/O. It is integrated with a small battery for data collection purposes only as the RFID cannot be employed in the actual processing pipes. The two-chip stack permits smaller imitation food particles than otherwise would be the case and is shown in Figure 1.5. The RFID coil can be seen. The chip includes capacitors for temporary power storage.

1.5 Cost Reduction

There have been several examples of 3DIC products with a focus on cost reduction, though often they achieve performance enhancement simultaneously.

Many cell phone cameras are today made as a two-chip stack. One chip is a backside-illuminated pixel array that does not include interconnect layers or even complete complementary metal–oxide–semiconductor (CMOS) transistors. The second chip is a complete CMOS chip on which the analog-to-digital converters (ADC) are built and interconnected for all other functionalities required of an image sensor. A high-density connection process such as hybrid bonding is used to create the needed face-to-face connections on a few micron pitch. Examples are reported in References 8 and 9. This is also an example of heterogeneous integration, as the two chips in the stack go through different manufacturing processes.

Another example of cost reduction is found in building high-end field-programmable gate arrays (FPGAs). To a first approximation, the cost of a large CMOS chip goes up with the square of the area. This is because the probability of a defect occurring on the chip and thus *killing* the chip goes up with the chip area, whereas the cost of making the chip in the first place also goes up with the area. Thus it is worth considering partitioning a large chip into a set of smaller ones, if the cost of integration and the additional test are less than the savings accrued to increase CMOS yield. Xilinx, California, United States, investigated this concept for large FPGAs and is now selling FPGA modules containing 2–4 CMOS FPGA chips, tightly integrated on an interposer. Details are not available but they claim an overall cost savings [10].

A third example is that of mixing technology nodes. In general, Moore's law tells us that a digital logic gate costs less to make in a more advanced technology due to the reduced area for that gate in that node. However, in contrast, many analog and analog-like functions such as ADC and high-speed serial deserializer I/Os (SerDes) do not benefit in such a fashion. The reason is that the analog behavior of a transistor has higher variation for smaller transistors than for larger ones. Thus for many analog functions that rely on well-matched behaviors of different transistors in the circuit, no benefit is accrued from building smaller transistors. More simply put, analog circuit blocks do not shrink in dimensions with the use of more advanced technologies. Thus the cost of these functions in a more advanced process node can actually be higher, than in the old node, as the old node costs less to make per unit of area. Again this is an example of heterogeneous integration, the heterogeneity being that of mixing technology notes.

Although Wu [10] also explored this concept generically, Erdmann et al. [11] have explored this concretely for a mixed ADC/FPGA design. Their design consisted of two 28 nm FPGA logic dies, integrated with two 65 nm ADC array dies on an interposer. Thus two sets of cost benefits are accrued, first the yield-related savings from splitting the logic die into two and the fabrication cost savings of keeping the ADCs in an older technology.

1.6 Heterogeneous Integration

In Section 1.5, two examples of heterogeneous integration were given in which wafers from different silicon process lines were integrated using 3DIC technologies. There is also considerable interest in integrating wafers with different underlying technologies, not just silicon.

Another interesting use of 3D technologies has been to build nonvisible light sensors, sometimes using a nonsilicon technology for the sensing layer. Examples include IR imagers, X-ray imagers [12], and other images for high-energy physics investigations [13].

The example is that of mixing III-V and silicon technologies. This is best exemplified by The Defense Advanced Research Projects Agency Diverse Accessible Heterogeneous Integration (DARPA DAHI) program in which GaN and InP chips are integrated on top of the CMOS chips through micro-bumps and other technologies [14]. More specifically, CMOS can be used for most of the transistors in a circuit, whereas GaN high-electron-mobility transistors (HEMTs) can be used for its high-power capability, and InP heterojunction bipolar transistors (HBTs) can be used for their very high speed. An example of the latter is an ADC. In an ADC, only a few transistors generally determine the sampling rate. Thus with the DAHI technology, these few transistors can be built in a high speed, but expensive and low yielding, InP chiplet, whereas the rest of the ADC is built in cheaper and more robust CMOS.

1.7 Performance Enhancement

A major goal of using 3D technologies is to increase the performance in some key aspects. For example, a two-layer image sensor might enable bigger pixels in the image layer and thus greater sensitivity. Alternatively, integrating a high-performance low device count InP chiplet onto a sophisticated CMOS die leads to performance improvement, often in conjunction with the high I/O count that the 3D technologies permit.

However, many 3D products and research objects are driven purely by performance, specifically the ability to integrate many silicon parts, with a lot of accumulated area into one 3D chip stack.

3D memories do exactly this and are positioned to be the next large-volume application of 3DIC technologies. To date dynamic random-access memory (DRAM) has relied on one-signal-per-pin signaling using low cost, low pin count, and single-chip plastic packaging. As a result, DRAM has continued to lag logic in terms of bandwidth potential and power efficiency. Furthermore, the I/O speed of one-signal-per-pin signaling schemes are unlikely to scale a lot beyond what can be achieved today in double data rate (DDR4) (up to 3.2 Gbps per pin) and graphics double data rate (GDDR6) (8 Gbps). Beyond these data rates, two-pins-per-signal differential signaling is needed. Furthermore, the I/O power consumption, measured as pJ/bit, is relatively high, even for the low-power DDR (LPDDR) standards (intended for mobile applications).

TABLE 1.1

Comparison of 2D and 3D Memories

Technology	Capacity	BW (GB/s)	Power (W)	Efficiency (mW/GB/s)	I/O Efficiency (mW/Gb/s)	DQ Count
DDR4-2667	4 GB	21.34	6.6	309	6.5–39	32
LPDDR4	4 GB	Up to 42	5.46	130	2.3	32
HMC	4 GB	128 GBps	11.08	86.5	10.8	8 Serdes lanes
HBM	16 GB	256 GBps		48		1024
Wide I/O	8–32 GB	51.2 GBps[a]		42[b]		256
DiRAM4	64 GB	8 Tbps				4096

[a] Wide I/O2, Wide I/O1 was half of this.
[b] Wide I/O1, Wide I/O2 should be lower.

There have been several proposed 3D memory technologies, which are summarized together with 2D memories in Table 1.1. Note that in this table, both B (Byte) and b (bits) are used. Also note that 1 mW/Gb/s is equivalent to 1 pJ/bit.

The Hybrid Memory Cube (HMC) is a joint Intel-Micron standard that centers on a 3D-stacked part including a logic layer and multiple DRAM layers organized as independent vertical slices. This 3D chip stack is then provided as a packaged part, so the customer does not have to deal with any 3DIC or 2.5D packaging issues. At the time of writing this chapter, Micron offered 2 GB and 4 GB parts with a maximum memory bandwidth of up to 160 GBps (The 128 GBps part is used in Table 1.2). The data I/O is organized as an eight high-speed serial channels or lanes. HMC is mainly aimed at computing applications. However, it does not appear to be widely adopted.

The high bandwidth memory (HBM) is a JEDEC (i.e., industry) standard that is not intended to be packaged on its own but to be further integrated using an interposer or 3D chip stack. Its high pin count, configuring 8 × 128-bit wide interfaces, prevents it from being packaged easily. The interfaces run at 2 Gbps per pin and the pins are placed on a dense 48 × 55 μm grid. It is fabricated as a stack of multibank memory die, connected to a logic

TABLE 1.2

Improvements in 3D Design over 2D Using Logic Cell Partitioning

	Total Wire Length (% Change)	F_{max} (% Change)	Total Power (% Change)	Power/MHz
Radar PE	−21.0%	+22.6%	−12.9%	−38%
Advanced encryption standard (AES)	−8%	+15.3%	−2.6%	−18%
Multiple-input and multiple-output (MIMO)	+21.6%	+17.1%	−5.1%	−23%

die through a TSV array, the TSV arrays running through the chip centers. Each chip is F2B mounted to the chip beneath it. The eight channels are operated independently. Details for a first-generation HBM (operating at 3.8 pJ/bit power level at 128 GB/s) can be found in Reference 15. The use of HBM in graphics module products has been announced by Nvidia and AMD.

Wide I/O is also a JEDEC-supported standard, aimed largely at low-power mobile processors. Although intended to be mounted on top of the logic die in a true 3D stack, side by integration on an interposer is also possible. Wide I/O is a DRAM-only stack—there is no logic layer. Instead the DRAM stack is exposed through a TSV-based interface and the memory controller is designed separately on the CPU/logic die that is customer designed. To date the thermal challenges of mounting a DRAM on an already hot mobile processor logic die have been insurmountable, especially as it is desired to operate the DRAM at a lower temperature than logic (85°C for DRAM vs. 105°C for logic) to control leakage and refresh time. This has been a barrier to adoption.

The Tezzaron DiRAM4 is a proprietary memory still in development. It has 4096 data I/O organized across 64 ports. It is intended only for 3D and interposer integration. It has a unique organization in that the logic layer is not only used for controller and I/O functions but also houses the global sense amplifiers and addresses decoders that in other 3D memories are on the DRAM layers. This permits faster operation for these circuits. The DiRAM4 has potential for a very high bandwidth (up to 8 Tbps) and fast random cycles (15 ns) [16].

1.8 Power Efficiency

In theory, taking a 2D chip and turning it into a stack of two chips should reduce the power of the chip by about 35%. The reason is quite simple. In most digital chips, about half (50%) the power is consumed in driving wires. When stacked, the two chips in the stack should be half of the area of the original chip, and thus the wires should be $\sqrt{2}$ shorter. As the power needed to drive a wire is roughly proportional to its length, each wire driver should see a power reduction of around 70%, that is, the chip as a whole should see a power reduction of around 35%.

However, this is difficult to achieve as short wires are unlikely to be routed between the two chips. On the other hand, reduced wire loads also means the chip can operate faster. Thus one would still expect to see a substantial improvement in the power/performance ratio over the 2D equivalent.

In one project at North Carolina State University, we demonstrated this using a two-chip CMOS stack-attached F2F with a copper thermo-compression-bonded interface at a 6.3 μm pitch. We put together

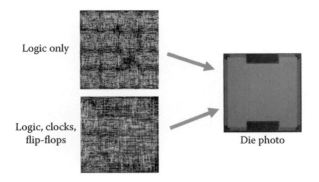

FIGURE 1.6
Two-chip stack.

a CAD flow to do this that could reuse 2D CAD tools, especially place and route tools. To make that feasible, all flip-flops are kept in one tier so that 3D clock distribution was not required. The radar PE was implemented in the Tezzaron bulk CMOS 3D process [17] (Figure 1.6). The results are summarized in Table 1.2. On average, performance per unit of power was increased by 22% due to the decreases in wire length achieved through this partitioning approach. The radar processor had an improvement in performance per unit of power of 21%. The other designs achieved 18% and 35%. The achieved improvement was roughly equivalent to one generation of Moore's law scaling.

In a different project at North Carolina State University, we took a very different approach to improving performance/power using 3D technologies. A stack of two different CPUs are integrated vertically using a vertical *thread transfer* bus that permits fast compute load migration from the high-performance CPU to and from the low-power CPU when an energy advantage is found [18]. In this design, the *high-performance CPU* can issue two instructions per cycle, whereas the *low-power CPU* is a single-issue CPU. The transfer is managed using a low-latency, self-testing multisynchronous bus [19]. The bus can transfer the state of the CPU in one clock cycle by using a wide interface and by exploiting a high-density copper–copper direct bond process. The caches are switched at the same time, removing the need for a cold cache restart.

Simulation with Specmark workloads shows a 25% improvement in the power/performance ratio compared with executing the sample workload solely in the high-performance processor. In contrast, if the workload was executed solely in the single-issue (*low-power*) CPU, there would be 28% total energy savings, compared with keeping the workload in the high-performance CPU but at the expense of a 39% reduction in performance. If the workload was allowed to switch every 10,000 cycles, there would be 27% total energy savings but at the expense of only a 7% reduction in performance that is, a 25% improvement in power per unit of performance is achieved. The die photograph is shown in Figure 1.7.

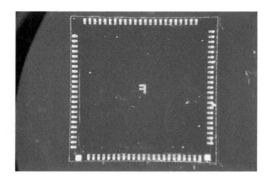

FIGURE 1.7
3D heterogeneous processor die.

1.9 Conclusion

3D and 2.5D technologies considerably open the design space for semiconductor technologies. Dimensions for exploration include miniaturization, cost reduction, achieving new modalities via heterogeneous integration, performance improvement, and improvement in performance/power.

References

1. P. Enquist, G. Fountain, C. Petteway, A. Hollingsworth, and H. Grady, Low cost of ownership scalable copper direct bond interconnect 3D IC technology for three dimensional integrated circuit applications, *IEEE International Conference on 3D System Integration, 3DIC 2009,* San Francisco, CA, 2009, pp. 1–6.
2. U.S. Patents 5,501,893 and 6,531,068.
3. J.A. Burns, B.F. Aull, C.K. Chen, C. Chang-Lee, C.L. Keast, J.M. Knecht, V. Suntharalingam, K. Warner, P.W. Wyatt, and D. Yost, A wafer-scale 3-D circuit integration technology, *IEEE Transactions on Electron Devices,* 52(10): 2507–2516, 2006.
4. http://image-sensors-world.blogspot.com/2008/09/toshiba-tsv-reverse-engineered.html
5. http://www.semicontaiwan.org/en/sites/semicontaiwan.org/files/docs/4._mkt__jerome__yole.pdf
6. G. Chen, M. Fojtik, D. Kim, D. Fick, J. Park, M. Seok, M.-T. Chen, Z. Foo, D. Sylvester, and D. Blaauw, Millimeter-scale nearly perpetual sensor system with stacked battery and solar cells, *2010 IEEE International Solid-State Circuits Conference-(ISSCC),* San Francisco, CA, 2010, pp. 288–289.
7. A. Lentiro, Low-density, ultralow-power and smart radio frequency telemetry sensor, PhD Dissertation, NCSU, 2013.

8. P. Enquist, 3D integration applications for low temperature direct bond technology, *2014 4th IEEE International Workshop on Low Temperature Bonding for 3D Integration (LTB-3D)*, Tokyo, Japan, 2014, p. 8.

9. http://www.sony.net/SonyInfo/News/Press/201201/12-009E/index.html, http://www.reuters.com/article/2014/03/25/us-sony-sensors-idUSBREA2O0PQ20140325

10. X. Wu, 3D-IC technologies and 3D FPGA, *3D Systems Integration Conference (3DIC), 2015 International*, Sendai, Japan, 2015, pp. KN1.1–KN1.4.

11. C. Erdmann et al., A heterogeneous 3D-IC consisting of two 28 nm FPGA die and 32 reconfigurable high-performance data converters, *IEEE Journal of Solid-State Circuits*, 50(1): 258–269, 2015.

12. G.W. Deptuch et al., Fully 3-D integrated pixel detectors for X-rays, *IEEE Transactions on Electron Devices*, 63(1): 205–214, 2016.

13. http://meroli.web.cern.ch/meroli/DesignMonoliticDetectorIC.html

14. S. Raman, C.L. Dohrman, and T.H. Chang, The DARPA diverse accessible heterogeneous integration (DAHI) program: Convergence of compound semiconductor devices and silicon-enabled architectures, *2012 IEEE International Symposium on Radio-Frequency Integration Technology (RFIT)*, Singapore, 2012, pp. 1–6.

15. D.U. Lee et al., A 1.2 V 8 Gb 8-channel 128 GB/s high-bandwidth memory (HBM) stacked DRAM with effective I/O test circuits, *IEEE Journal of Solid-State Circuits*, 50(1): 191–203, 2015.

16. www.tezzaron.com and Evolving 2.5D and 3D integration, RTI 3D ASIP, December 2013.

17. T. Thorolfsson, S. Lipa, and P.D. Franzon, A 10.35 mW/GFLOP stacked SAR DSP unit using fine-grain partitioned 3D integration, *Proceedings in Custom Integrated Circuits Conference 2012*, pp. 1–4.

18. E. Rotenberg, B.H. Dwiel, E. Forbes, Z. Zhang, R. Widialaksono, R. Basu Roy Chowdhury, N. Tshibangu, S. Lipa, W.R. Davis, and P.D. Franzon, Rationale for a 3D heterogeneous multi-core processor, *2013 IEEE 31st International Conference on Computer Design (ICCD)*, pp. 154, 168, October 6–9, 2013.

19. Z. Zhang, B. Noia, K. Chakraparthy, and P. Franzon, Face to face bus design with built-in self-test in 3DICs, *Proceedings IEEE 3D Integration Conference*.

2

Three-Dimensional System-in-Package for Application-Specific Integrated Circuit and Three-Dimensional Dynamic Random-Access Memory Integration

Li Li

CONTENTS

■■■

2.1 Three-Dimensional SiP Introduction

The bandwidth for high-performance networking switches and routers increases two to ten times for every new generation. This in turn drives the bandwidth requirements for the application-specific integrated circuits (ASICs) and their external memory devices designed for the high-performance network systems. To meet these bandwidth requirements, the ASIC packaging technology has to keep up with the integrated circuit (IC) technology scaling. Recently several 2.5-dimensional (2.5D) and 3-dimensional (3D) IC integration or packaging technology platforms have been developed to address the gap seen between the slowdown of Moore's law scaling and the ever-increasing system integration requirements. In the remainder of this chapter, the terms 2.5D and 3D IC integration and 2.5D and 3D packaging are used interchangeably.

The early success of FCAMP that stands for Flip Chip and Memory Package provided a much-needed solution to the bandwidth challenges between the ASIC and memory devices designed for a high-end networking application [1]. The FCAMP, sometimes also referred to as a system-in-package (SiP), contains an ASIC die that is attached to an organic substrate using the flip chip technology. The memory devices are packaged, tested at-speed and burned-in before putting on the SiP substrate. This process flow is compatible to the now industry-standard flip chip assembly process and has achieved high manufacturing yield. The other innovation that made the FCAMP a success is the availability of large size (>52.5 mm), high density, thin-core or coreless, and high-performance organic substrates that are based on the build-up, micro-via technologies [2]. The bandwidth improvement through the FCAMP design has had a profound impact on high-performance networking switches and routers, similar to computing applications.

Leveraging 3D IC integration, a concept of 3D ASIC and memory integration with a silicon interposer and through-silicon via (TSV) was proposed and a feasibility study was conducted [3]. 3D ASIC and memory heterogeneous integration can be considered as an extension of the FCAMP that was developed about a decade ago. It has the advantages of reducing power consumption, improving the bandwidth between the ASIC and memory, and modularizing system hardware designs. Critical components for enabling 3D ASIC and memory integration include large-size silicon interposer, 3D-stacked memory with TSV, and microbump or micropillar interconnects.

The 3D ASIC and memory heterogeneous integration not only optimizes the package's size but also imposes certain thermal limitations, as heat generated from the bottom logic chip has to pass through the entire die stack to get dissipated. To overcome the challenges associated with 3D IC integration; especially in the areas of generating TSVs in the active devices, thermal management and packaging; a series of multichip packaging technologies using a silicon interposer or a fine-pitch organic interposer have been developed [4]. This type of IC integration is nicknamed 2.5D IC integration as the chips are still packaged in a planar format on the interposer as seen in previous generations of multichip module (MCM) technologies. Furthermore, a 3D SiP combining 2.5D and 3D IC integration has been developed for integrating one or more ASIC chips with one or more 3D-stacked memory die stacks. Both the ASIC chip and 3D memory die stacks reside on one side of the interposer [5]. A schematic of the 3D SiP is shown in Figure 2.1.

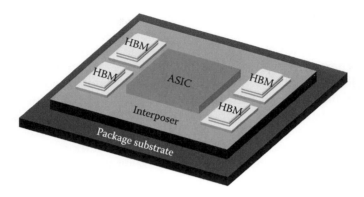

FIGURE 2.1
A schematic of a 3D SiP with one ASIC die and four HBM die stacks.

Here HBM stands for high bandwidth memory, which is a 3D-stacked dynamic random-access memory (DRAM) device developed by major DRAM suppliers in recent years.

The interposers used in 2.5D or 3D SiP can provide much higher wiring density than the conventional organic build-up substrate. Together with the microbump interconnect technology, the 3D ASIC and memory SiP can greatly increase the width and density of the interface between the ASIC and the memory devices, and hence the bandwidth. The 3D SiP for ASIC and memory integration has the following advantages:

- High package thermal performance similar to the single flip-chip module
- Higher interconnect wiring density than the conventional organic build-up substrate
- Compatible to the flip-chip assembly process
- Lower CTE mismatch between the silicon chips and the interposer
- Lower stresses on the low-k dielectrics and interconnects of the silicon chips by leveraging the Cu-filled interposer as a *stress buffer*

2.2 Enabling Technologies for 3D SiP

2.2.1 Three-Dimensional Stackable Memory

In the past several years, memory suppliers started the development of low-power, high-bandwidth, 3D-stackable memory devices. Industry standards for the 3D-stackable memory devices have also been developed. An early example of 3D IC integration is the application of 3D-stacked (3DS) DRAM with TSV [6]. Published in December 2013, the addendum to JESD79-3 defines the 3DS DDR3 SDRAM specification, including features, functionalities, AC and DC characteristics, packages, and ball/signal assignments. Since then, 3DS DDR3 and 3DS DDR4 have been developed and are now in mass production to meet the ever-increasing demand in memory module density for server, high-performance computing (HPC) and networking applications. In parallel to the development of 3DS DDR3 and 3DS DDR4, Wide I/O mobile DRAM was developed leveraging the same 3D IC stacking technology. The Wide I/O single date rate (SDR), which is 512-bit wide, running at 200 MHz can provide a total bandwidth of 12.8 GB/s and consumes only half the power of a low-power double data rate (LPDDR2) on the per bandwidth base [7]. The second-generation Wide I/O DRAM, Wide I/O 2, provides four times the memory bandwidth (up to 68 GB/s) of the previous version but at lower power consumption [8]. From a packaging perspective, the Wide I/O DRAM is designed specifically to stack on top of a system-on-chip (SoC) logic device such as an application processor (AP) die using TSVs and microbumps or micropillars to minimize electrical interference and the overall package form factor that is much needed for mobile applications.

Using a similar 3D heterogeneous integration approach, a consortium led by Micron Technology developed the Hybrid Memory Cube (HMC) [9]. The HMC consists of a 3D stack of DRAM dies on top of a logic memory interface/controller die. The DRAM dies within the stack are each of 1 GB in DRAM density and are divided into 16 partitions. Each partition has two banks (also called arrays), and its own data and control TSVs. HMC is designed to emphasize massive amounts of bandwidth at higher power consumption and cost than Wide I/O 2.

For the 3D ASIC and memory SiP designs, these 3D-stackable memory devices have advantages that include the following:

- Memory and ASIC (logic) device can each be built in their own specific processes.
- Further exploit the process/cost differences between the logic and memory devices.
- Very high data rate (bandwidth) with low latency and low-power per bandwidth.
- Wide interface enabled by very wide interdie interconnect.
- Low parasitic enabled by short, direct interconnect.

Recently, this SiP approach has been extended to include the HBM DRAM devices based on 3D IC integration with TSV and micropillar interconnects [5].

HBM is a new class of DRAM developed by major DRAM suppliers leveraging wide I/O and TSV technologies [10]. It is targeted for graphics processor units (GPU), HPC, server, networking, and client applications where peak bandwidth, bandwidth per watt, and capacity per area are valued metrics. The high bandwidth is delivered using a 1024-bit wide memory interface that is divided into eight independent channels. To accommodate the 1024-bit interface between HBM and the host compute die, micropillars with a pitch of 55 microns are placed at the bottom of the HBM DRAM die stack.

2.2.2 High-Density Interposer

2.2.2.1 Silicon Interposer

The introduction of 3D IC integration also brought new requirements and sometimes disruptions to the existing microelectronics manufacturing supply chain. One example is the manufacturing and supply of silicon interposers (Si-IP). Currently there are three commonly used interposer-manufacturing process flows. The first can be referred to as the *Foundry Process* flow. The silicon interposer is fabricated completely by the wafer foundry and sometimes the service even includes packaging the 3D IC subassembly to the package substrate [11].

The second can be referred to as the *middle-end-of-line (MEOL) process* flow. MEOL is often used by the outsourced assembly and test (OSAT) suppliers for processing TSV wafers with active circuits. It is usually started at the wafer foundry for via generation and filling, and fabrication of front-side metal wiring layers. Then the full thickness wafers are delivered to the OSAT for further downstream processes that include wafer thinning, TSV reveal, backside metal layer generation, passivation, and bumping. The third process can be called as the *substrate process* flow. In this case, the silicon interposer is fabricated and supplied by the traditional packaging substrate suppliers. A comparison of these silicon interposer manufacturing and supply chain flows is given schematically in Figure 2.2.

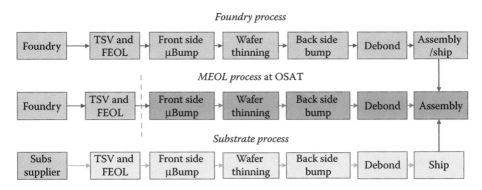

FIGURE 2.2
A comparison of silicon interposer manufacturing and supply chain flows.

For both the foundry and MEOL processes, TSVs are generated using the deep reactive-ion etching (DRIE) process [12]. The front-side interconnects or wiring layers are made with Cu damascene techniques. For the backside interconnection, MEOL usually uses redistribution layer (RDL) process [13]. For true 3D wafers with active circuits and TSVs, MEOL may be preferred but for passive interposers, an alternative and cost-effective way may exist. This alternative, *substrate process* flow is the focus of this study.

The 3D SiP in Figure 2.1 can be supported with a silicon interposer with TSV. The TSVs that are typically 10–25 µm in diameter are formed by the DRIE process. The walls of the TSV are lined with the SiO_x dielectric. Then, a diffusion barrier and a copper seed layer are introduced. The via holes are filled with copper through the electrochemical deposition process. The chemical–mechanical planarization (CMP) process is used to remove the copper overburden.

Recently, manufacturing of cost- and performance-effective, large-size silicon interposer has been investigated [14]. The existing supply chain and infrastructure of high-performance flip-chip packaging substrates are leveraged. There are several advantages in this approach. One is minimal disruption to the existing supply chain. The silicon interposer is considered as a packaging material rather than another piece of silicon chip. Secondly, large-size silicon interposers can be manufactured with a line width and line spacing in the range of a few micrometers. This type of silicon interposer (Si-IP) is often referred to as coarse-pitch silicon interposer to distinguish itself from the ones made by wafer foundries. Table 2.1 shows a comparison between the coarse-pitch silicon interposer to the fine-pitch silicon interposer.

For coarse-pitch silicon interposers, the semi-additive process (SAP) is used to fabricate Cu wiring on either side of the interposer. The SAP method

TABLE 2.1

Comparison of Coarse-Pitch and Fine-Pitch Silicon Interposers

Features	Coarse-Pitch Si-IP	Fine-Pitch Si-IP
Cu wiring (dielectric)	SAP (polyimide)	Damascene (oxide)
Microbump material	Cu/Ni/SnAg, etc.	Cu/Ni/SnAg, etc.
Minimum microbump size/pitch (µm)	30/50	20/40
Minimum front-side wiring Line width/space/thickness (µm)	3/3/3	0.5/0.5/1.0
RDL via size (µm)	20	1.0
TSV size/pitch/depth (µm)	60/150/200	10/50/100
Minimum back-side wiring Line width/space/thickness (µm)	10/10/3	10/10/1.0
Back-side pad and bump size (µm)	Ni/Au 100/150	Ni/Au 100/150

FIGURE 2.3
A top view of a 35 × 35 mm silicon interposer attached to a package substrate.

has fewer process steps and uses conventional equipment that is also used for fine-pitch printed wiring board fabrication [13]. On the other hand, fine-pitch silicon interposer relies on the damascene technique for Cu wiring fabrication that requires both chemical–mechanical polishing (CMP) and dry etching processes. As it involves fewer process steps and uses conventional equipment for fabrication, coarse-pitch silicon interposers will be less costly when compared to fine-pitch silicon interposers [13].

Figure 2.3 shows the top view of a 35 × 35 mm silicon interposer fabricated with the SAP method.

The front side of the interposer has two metal wiring layers, whereas the backside has one wiring layer. The interposer shown is attached on a 50 × 50 mm HiTCE ceramic substrate. Major fabrication steps used are shown schematically in Figure 2.4.

The size of the silicon interposers from the leading foundries is currently limited to 26 × 32 mm, which is the reticle size used in the lithographic wafer processing. This size limitation can be a disadvantage for ASIC and memory integration as the die sizes for high-performance ASICs can be as large as 25 × 25 mm and the ASIC chips require multiple external memory devices as illustrated in the FCAMP case. Use of reticle stitching to increase the silicon interposer size is under development; however, it has its own limitations as well.

2.2.2.2 Organic Interposer

Another approach to overcome the size limitation of the silicon interposer technology and for better ASIC and memory integration is through an

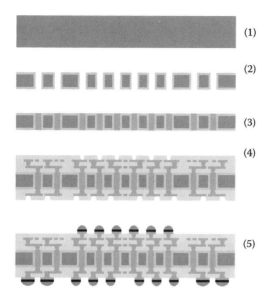

FIGURE 2.4
Major fabrication steps used to fabricate the coarse-pitch silicon interposer. (1) Silicon wafer; (2) TSV formation, thermal oxidation silicon thickness: 200 μm, TSV: diameter 60 μm/pitch 150 μm; (3) TSV filling, planarization; (4) multilayer wiring (double-sided), RDL—semi-additive process, insulator—photosensitive resin, top side—two-layer, and bottom side—one-layer; and (5) double-sided bumping, Cu/Ni/SnAg bump (electroplating), and diameter—30 μm.

organic interposer technology. A new 3D SiP based on an organic interposer is being developed [5]. In this study, application of a fine-line, fine-pitch, large-size organic interposer for integrating ASIC and HBM is investigated. There are several advantages of organic interposers. One is less disruption to the existing supply chain while leveraging the infrastructure developed over the years for manufacturing high-performance flip-chip packaging substrates. The organic interposer is considered as a packaging material rather than another piece of silicon chip for the 3D SiP assembly. Second, large-size organic interposers can be manufactured with a panel process and can have line width and line spacing in the range of a few micrometers making them suitable for integrating large die-size ASIC or other logic devices with multiple HBM DRAM stacks. Table 2.2 shows a comparison between the organic interposer and a typical fine-pitch silicon interposer.

To accommodate a total of 1024 wires and micropillars for the HBM interface, a few more routing layers are needed for the organic interposer due to its coarse line width and spacing compared to the silicon interposer. Major manufacturing steps for making organic interposers are the same as that for the organic build-up substrates. These include the following:

TABLE 2.2

A Comparison of Organic Interposer and Silicon Interposer

Features	Organic Interposer	Silicon Interposer
Cu wiring (dielectric)	SAP pattern plating (organic)	Damascene (oxide)
Minimum front-side pad size/pitch (μm)	30/55	20/40
Minimum front-side wiring Line width/space/thickness (μm)	6/6/10	0.5/0.5/1.0
Number of routing layers (typical)	10	4
Wiring layer via size (μm)	20	1.0
PTH or TSV size/pitch/depth (μm)	57/150/200	10/50/100
Back-side pad size/pitch (μm)	100/150	100/150
Minimum back-side wiring Line width/space/thickness (μm)	6/6/10	10/10/1.0

- Plated-through hole (PTH) generation and filling for the core layer
- Circuitization of the core layer
- Building Cu-wiring layers on two sides of the core layer with the micro-via and build-up processes

Pattern plating as part of the SAP is used to fabricate Cu wiring and micro-vias for all the build-up layers. The pattern plating method has been used extensively in manufacturing high-density, build-up organic packaging substrates. On the other hand, fine-pitch silicon interposer relies on the damascene technique for Cu-wiring fabrication that requires both chemical–mechanical polishing (CMP) and dry etching processes. As it involves fewer process steps and uses a panel format for fabrication, organic interposers will be less costly when compared to silicon interposers and can offer much larger size interposer for high-performance ASIC and HBM integration.

Figure 2.5 shows the top view of a 38 × 30 mm organic interposer fabricated with the build-up and pattern-plating processes.

As shown in the above-mentioned figure, four micropillar footprint patterns are included for attaching the HBM DRAM die-stacks. A close-up view on the pads for micropillars is shown in Figure 2.6.

2.2.3 Microbump Interconnect

To accommodate the higher wiring density enabled by silicon or organic interposers and to join 3D-stackable memory devices, the pitch and size of conventional controlled collapsed chip connection (C4) solder bumps have to be reduced greatly. Based on the same electroplating process developed for C4 solder bumps, a microbump interconnect technology is developed and applied to a 3D SiP test vehicle [3,5]. A 3D view of the electroplated micro-bumps is shown in Reference 3. The bump shear test is used to characterize

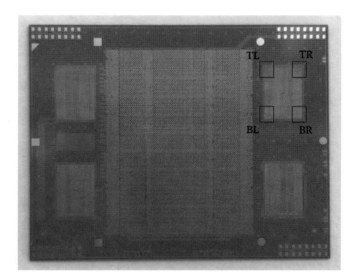

FIGURE 2.5
A top view of a 38 × 30 mm organic interposer manufactured.

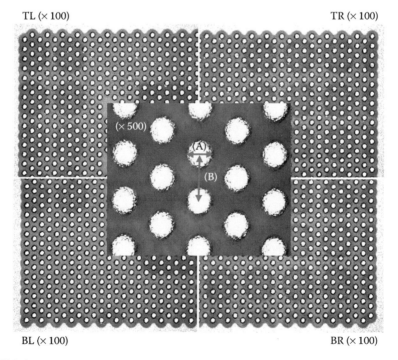

FIGURE 2.6
A close-up view of a 38 × 30 mm organic interposer on the pads for micro-pillars.

the mechanical integrity of the microbumps after the plating process and after aging at 150°C for 72 hours. The results for the bump shear test before aging are also shown in Reference 3. The average shear force is 3.47 g/bump. After aging, the average shear force decreased to 2.1 g/bump and the growth of intermetallic compound (IMC) Cu_6Sn_5 is observed [3]. Another approach for the solder bump pitch reduction and density increase is through a copper (Cu) pillar approach. The Cu pillar technology was first introduced to advanced logic devices in 2006 and has since been developed especially for flip chip applications with a bump pitch below 100 μm. Copper pillars are less prone to electromigration that makes the technology a good choice for applications with reduced bump pitches, sizes, and increased current densities. The Cu pillar interconnection used for 2.5D and 3D IC integration is also referred to as micropillar interconnection. For example, the micropillar array used to connect the current generation of HBM DRAM to interposers has a pitch of 55 μm [15].

2.3 3D SiP for Application-Specific Integrated Circuits and High Bandwidth Memory Integration

A 3D SiP is designed and manufactured that includes a large-size organic interposer with five Cu wiring layers on each side of the core layer. The organic interposer has a size of 38 × 30 × 0.4 mm. A high-performance ASIC die measured in 19.1 × 24 × 0.75 mm is attached on top of the organic interposer along with four HBM DRAM die stacks. The HBM DRAM die stack with a size of 5.5 × 7.7 × 0.48 mm includes one base buffer die and four DRAM-core dice, which are interconnected with TSVs and fine-pitch micropillars. The HBM stacks are placed equally in quantity along the two sides of the ASIC die. For assessing thermomechanical reliability, two of the HBM stacks are replaced with their daisy-chain versions.

The ASIC and the HBM 3D IC subassembly is then assembled on an organic packaging substrate with the conventional C4 solder bumps. Communications between the ASIC die and the HBM DRAM stacks are made through the Cu wiring and micro-via interconnections of the organic interposer. A top view of the 3D SiP designed is shown schematically in Figures 2.7 and 2.8 provides a schematic cross-sectional view of the assembly.

Here *HBM-M* represents the mechanical or daisy-chain version of the HBM DRAM stack.

The 3D SiP was designed and built to provide functions and capabilities to test the signal and power integrity of the signal connections, of the HBM memory channels, from the ASIC die to the two HBM DRAM die stacks.

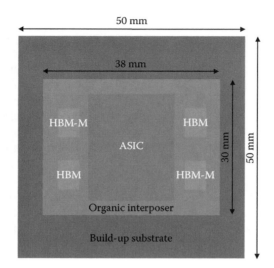

FIGURE 2.7
A schematic top view of the 3D SiP designed.

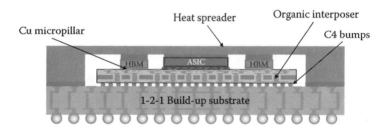

FIGURE 2.8
A schematic cross-sectional view of the 3D SiP designed.

A memory controller and PHY for the HBM were designed and implemented for the host ASIC on the organic interposer. To shorten the development cycle, the ASIC design including the IP for the HBM memory controller and PHY was first implemented through a field-programmable gate arrays (FPGA) device.

2.3.1 Organic Interposer Design

The 38 × 30 mm organic interposer has 12 layers: 5 top routing layers, 2 layers around the core, and 5 bottom routing layers. This configuration is also known as a 5-2-5 layer stack-up. The 50 × 50 mm package substrate has 4 layers: 1 top routing layer, 2 layers around the core, and 1 bottom routing layer. This configuration is known as a 1-2-1 layer stack-up.

Although the package substrate uses regular dielectric material used for standard build-up substrates, the organic interposer is designed and fabricated with a low-loss dielectric material (0.005@10 GHz loss tangent) and has a coefficient of thermal expansion (CTE) closely matched to the CTE of the dielectric material used in the packaging substrate. The substrate has an 800 μm thick core to help minimizing the warpage effects during reflow, whereas the organic interposer has a 200 μm thick core. This new, low-loss dielectric material used for the organic interposer allows ultrafine line spacing (line width/spacing = 6 μm/6 μm), low transmission loss, and high insulation reliability.

2.3.2 Simulation and Results

Time-domain and frequency-domain simulations have been performed to validate the 3D SiP design and to check if the performance specification of the HBM channels is met. The frequency-domain simulations have been done with the Cadence Sigrity PowerSI, whereas for the time-domain simulations, HSPICE was used.

The results for the frequency-domain analysis are shown in Figures 2.9 through 2.11, respectively.

Very good margins, based on standard specifications, for the 0–1 GHz operating frequency bandwidth are achieved. The insertion loss is −0.5 dB@1 GHz (with −1 dB@1 GHz standard specification), the return loss is approximately −23 dB@1 GHz (with −10 dB@1 GHz standard specification), and the cross-talk is approximately −45 dB@1 GHz (with −30 dB@1 GHz standard specification). The results in these figures also show that in fact good performances are valid for a very high-frequency bandwidth, up to 20 GHz.

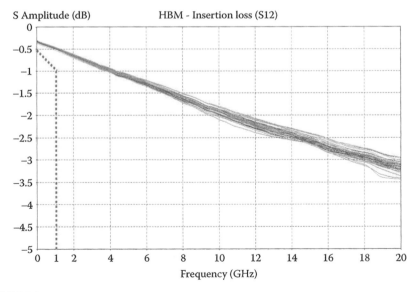

FIGURE 2.9
Insertion loss of the HBM channel simulated.

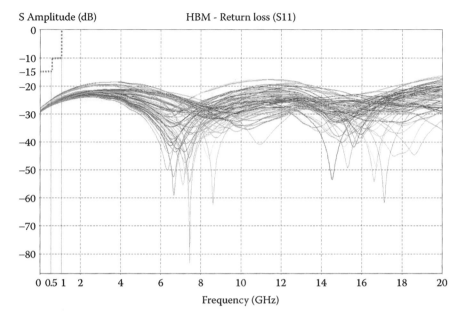

FIGURE 2.10
Return loss of the HBM channel simulated.

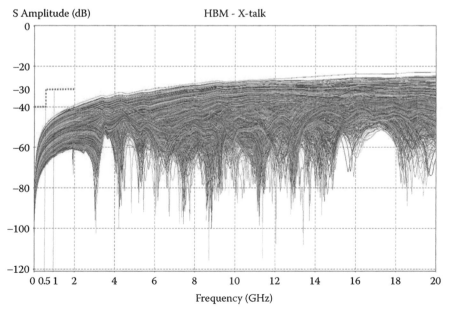

FIGURE 2.11
Loss due to cross-talk for the HBM channel simulated.

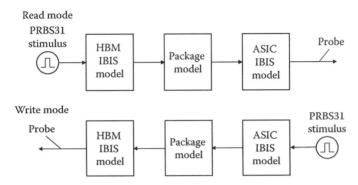

FIGURE 2.12
A schematic for setting up the time-domain analysis.

For the time-domain analysis, multiple corners have been simulated and only the worst-case results are reported here. From the package design file, an HSPICE model was extracted and used for the simulations, along with I/O buffers for the ASIC and HBM memories. The setup used in the HSPICE simulations is shown schematically in Figure 2.12.

A random PRBS 31 (pseudo-random bit sequence) is generated at the input and comparing the resulted sequence at the output with and without any aggressors (one signal was considered victim, and three aggressors on each side) is used to analyze the effect of cross-talk. The results are a comparison for the victim, in the situation with and without the aggressors, so that we can see the output data waveforms when the memory has low activity compared to the situation when the memory has high activity. As such, in READ MODE (see block diagram in Figure 2.12), the worst-case skew due to cross-talk is ~46 ps, whereas in WRITE MODE, the worst-case skew is ~42 ps.

Eye diagrams have been computed as well for the two modes. In READ MODE, with cross-talk from aggressors, the eye has a height of ~900 mV and a width of ~800 ps, and the introduced jitter is about 200 ps. In WRITE MODE, with cross-talk from aggressors, the eye has a height of ~1 V and a width of ~825 ps, whereas the introduced jitter is about 175 ps. All the results mentioned here are with cross-talk and in the worst-case possible. As it can be expected, the results without the cross-talk, in the worst-case, are slightly better. Good confirmation to the frequency-domain results is achieved.

2.4 Three-Dimensional SiP Assembly

To develop the optimal chip joining process for attaching HBM die stacks with micropillars and an overall packaging assembly process flow for the final 3D SiP module, shadow moiré method was used to analyze the

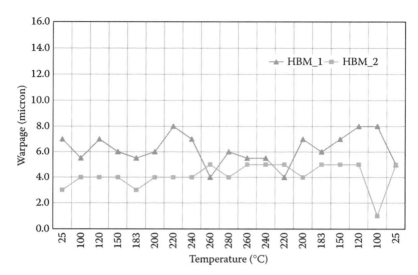

FIGURE 2.13
Warpages measured for the HBM die stack from room temperature to 280°C.

thermomechanical behavior of each constituent component of the 3D SiP. Figure 2.13 shows the warpage measurement results for the HBM die stacks over the temperature range from room temperature to 280°C.

As expected, the warpage of the HBM die stack was small (<8 μm) and did not change much with temperature.

Figure 2.14 shows the warpage measurement results for the organic interposer from room temperature to 260°C.

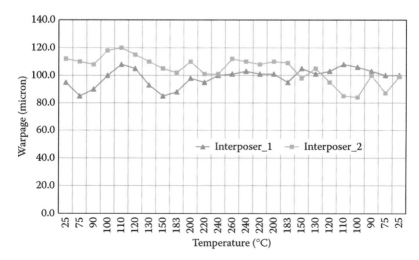

FIGURE 2.14
Warpages measured for the organic interposer from room temperature to 260°C.

It can be seen that the warpage of the organic interposer was relatively small (about 100 µm) and stayed in that range when the temperature was increased up to 260°C.

Based on the thermal deformation analysis for each components of the 3D SiP, a series of design-of-experiments (DOE) were planned and conducted to overcome the challenges in developing the suitable assembly process for the ASIC and HBM 3D SiP. As the warpages of the silicon dice and organic interposer were small and did not change with temperature, the ASIC die and HBM DRAM die stacks were assembled onto the interposer to form the ASIC and HBM subassembly first. Underfill encapsulation was used to protect the joints made of micropillars and regular bumps. A top view and a bottom view of the 3D SiP subassembly are shown in Reference 5.

The structure of the 3D SiP subassembly was then analyzed using X-ray, acoustic, and optical microscopy to ensure good micropillar solder joints are formed. Pictures of the cross-sections for the micropillar joints for attaching HBM die stacks can be found in Reference 5.

In the final assembly process step, the ASIC and HBM subassembly was attached to the package substrate using the conventional C4 solder bump interconnection. The C4 bumps were then encapsulated using the underfill material. A top view of the finished 3D SiP (without the lid) is shown in Figure 2.15 and a cross-sectional view of the finished 3D SiP (with the lid) is shown in Figure 2.16.

FIGURE 2.15
A top view of the finished 3D SiP.

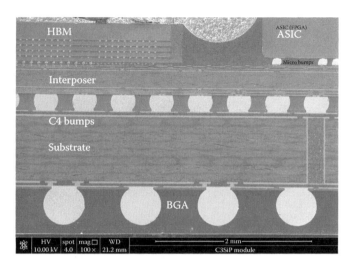

FIGURE 2.16
A cross-sectional view of the finished 3D SiP.

2.5 Test and Characterization

Electrical performance of the ASIC-HBM memory interface and functionalities of the HBM DRAM stacks as well as post-assembly *open and short* performance were verified and tested using an application evaluation board designed and fabricated.

The application board with a socket to host the 3D SiP modules was included as part of the test platform. The application board along with a 3D SiP placed in the test socket is shown in Figure 2.17.

The application board measures at 9.5 × 7 inches and has 16 layers: 7 ground layers, 3 power layers and 4 signal layers, and 2 high-speed layers, on the top and bottom of the printed circuit board (PCB). Although the top and bottom layers use Nelco N4000-13 EPSI dielectric material, the other layers use FR4 as dielectric.

Among the communication interfaces present on this application board, an I²C interface was included to provide the communication path from a Windows PC to the device and to output the results of the tests. The I²C interface was selected due to reuse of the memory controller built-in tests that also included an I²C interface. On the application board, test points and a connector are also included for testing the daisy-chain connections designed into the organic interposer, JTAG connection for configuring the ASIC, HBM direct access connector, and power distribution and clocking necessary for the operation of the ASIC and the HBM dies. On the PC, a Perl script-driven interface allows for the configuring, running, and analyzing of several

FIGURE 2.17
A top view of the application board with a 3D SiP module in the test socket.

different built-in memory tests within the ASIC memory controller design. The Perl script calls a Windows program and the necessary drivers that control a USB to I²C protocol conversion board that sits between the Application Board's I²C interface and the PC's USB port. Through the ASIC's built-in tests and via the I²C/Perl script platform, the electrical interface between PHY and memory is tested for general connectivity, cross-talk between lines, and simultaneously switching data lines. Defective cells in the memory are also detected through the tests.

2.6 Reliability Challenge

Compared with traditional 2D IC packaging, the emerging 2.5D and 3D IC integration involves several new elements in design, manufacturing, and supply chain processes. The traditional boundaries between the wafer fab and an OSAT supplier are blurred. These new elements include

- Thin die.
- TSV.
- Chip-to-chip (C2C) or chip-to-wafer (C2W) joining and interconnect.
- Micropillar or microbump for C2C or C2W connection.
- Die/interposer backside (BS) RDLs.

A summary of the new elements introduced with the 2.5D and 3D IC integration technologies and their implications on reliability of the final products is included in Table 2.3.

TABLE 2.3

New Elements Introduced by 2.5D and 3D IC Integration

Element	Failure Mechanism
Thin die	• Warpage-induced stress effect on device performance • Interconnect/Si delamination-cracking and misalignment • Deterioration of device characteristics by Cu diffused from the back side
TSV	• Si cracks and interconnect cracks/delamination • TSV-stress effects on device performance • Electromigration (EM)/stress migration (SM)
C2C connection	• Stress generated by chip-stacking effects device performance • Delamination/cracking in the interconnect
BS–RDL	• Stress-induced cracking/delamination • Cu contamination on the back surface of the thinned wafer/stress-enhanced diffusion
Microbumps	• Cracking, EM/SM, and electrical resistance
Large volume of copper	• Cu contamination on the back surface of the thinned wafer/stress-enhanced diffusion
Increased power density and large thermal gradients	• Device performance • Thermal runaway
Removal of I/O structure	• Electrostatic discharge (ESD) damage

The JEDEC publication, JEP158 titled *"3D Chip Stack with Through-Silicon Vias (TSVS): Identifying, Evaluating and Understanding Reliability Interactions"* [16] was developed about a decade ago when industry started to develop 2.5D and 3D IC integration for commercial applications. It was intended as a guideline to describe the extension of the standard tests to 3D IC components. The main element of this extension is the addition of appropriate test structures to evaluate the reliability of the TSVs and other new features introduced in the fabrication of 3D IC products. JEP158 together with JEDEC standard JESD47, *"Stress-Test-Driven Qualification of Integrated Circuits"* [17] are used as a starting point for technology qualification to assess the intrinsic reliability of TSV and micropillar interconnection that are new elements introduced by 2.5D and 3D IC integration.

Besides component-level reliability evaluation leveraging JEP158 and JESD 47, system-level and board-level validations are also needed. The following is an example of board-level reliability evaluation of a 3D IC component for networking applications. A test board was designed to mimic the actual system board. Four large body-size, high-performance Cisco ASIC packages were also included for the reliability test board assembly along with the two 3D IC components. Figure 2.18 shows a top view of the fully assembled test board assembly.

FIGURE 2.18
A top view of the reliability test board assembly for the ASIC and 3D IC components.

The 3D IC component consists of five dice vertically stacked and interconnected with TSV and microbumps. Over 3000 test features including chains going through all layers of the die stack and package substrate are designed into the 3D IC components. Two stress conditions were used. To assess the reliability of micropillars further, an enhanced stress condition was applied. It included 1000 hours of thermal aging at 125°C and followed by 6000 cycles of the temperature cycling with a range of 0°C–100°C. The stress conditions and final results are summarized in Table 2.4.

For the 3D SiP with ASIC and HBM components, a reliability test board was also designed, fabricated, and assembled [18]. A top view of the finished test board assembly designed for further board-level reliability evaluation is shown in Figure 2.19.

TABLE 2.4

Board-Level Reliability Evaluation of a 3D IC Package

Test	Condition	Sample Size	Results
Aging	1000 hours at 125°C	25	Zero failure
Aging plus temperature cycling	1000 hours at 125°C; 6000 cycles 0°C–100°C	25	Zero failure

FIGURE 2.19
A top view of the reliability test board assembly for the 3D SiP with ASIC and HBM DRAM.

Board-level temperature cycling test was conducted using an ESPEC TCC-150 environmental testing chamber, and real-time continuity resistance monitoring was done using an ESPEC AMR data logger. Daisy chains that cover the micropillar connection from the HBM stacks to the organic interposer, the regular solder bump connection from the organic interposer to the SiP substrate, and finally the ball grid array (BGA) connections to the test board are used as the test features for the reliability evaluation.

The reliability testing board measures at 10×8 inches and has 16 layers: 6 ground layers, 2 power layers and 6 signal layers, plus 2 layers on the top and bottom surfaces of the PCB. A front view showing the wiring and setup of the test boards inside the chamber is given in Figure 2.20.

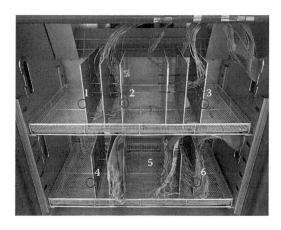

FIGURE 2.20
A front view showing the wiring and setup of test boards inside the environmental chamber.

A 0°C–100°C accelerated temperature cycling test was used for the board-level reliability test. The temperature profile inside the chamber was adjusted to minimize temperature variation from board to board. A measured temperature profile is given in Reference 18. The duration for the temperature cycling was 6000 cycles. Except for a few very early fails on some chains in the sub 100 cycle range, all the other daisy chains or test features monitored passed 6000 cycles with no failures.

2.7 Summary

To meet the requirements of next-generation Information and Communication Technology (ICT) systems, several 2.5D and 3D IC integration or packaging technology platforms have been developed. 3D-stackable memory, large-size silicon/organic interposers, and microbump interconnection are key-enabling technologies for realizing 2.5D and 3D IC integration. Leveraging the existing supply chain and infrastructure of high-performance flip-chip packaging substrates, a 3D SiP is designed and manufactured that includes a large-size organic interposer with a total of 12 Cu-wiring layers, an ASIC die, and 4 HBM DRAM stacks.

- The large-size organic interposer is manufactured with the SAP method established for the micro-via and build-up substrates, so it can be cost- and performance-effective for the applications such as integrating high-performance ASIC and HBM DRAM stacks.
- Thermal deformation of each constituent component of the 3D SiP was analyzed using real-time, metrological techniques such as shadow moiré to optimize material selection, assembly process, and reliability.
- A series of DOE and structural analyses were conducted to demonstrate that the 3D SiP module can be built with micropillar interconnects, a unique C2C joining, and the final packaging assembly processes developed.
- Electrical performance of the ASIC-HBM memory interface and functionalities of the ASIC and HBM DRAM stacks as well as the *open and short* of critical interconnects were verified and tested using an application evaluation board fabricated.

Understanding the new elements introduced by 2.5D and 3D IC integration and assessing their impact on reliability of the 2.5D and 3D devices are key to successfully introducing these new packaging technologies to server, HPC, and networking applications.

New elements introduced by 2.5D and 3D IC integration are reviewed and their impact on product reliability is discussed.

- The stress-test-driven methodology outlined by JEDEC standard JESD 47 and JEP157 is a good start point for evaluating reliability of 2.5D and 3D IC components and for assessing the individual and collective effects of the new elements introduced by 2.5D and 3D IC integration.
- Board-level or system-level characterization with purposely designed test vehicles and actual 3D IC components is used to validate reliability of 2.5D and 3D IC packages. Examples with one 3D IC component and one 3D SiP with ASIC and HBM DRAM were given as case studies.

A lot of progress has been made in the last decade through technology development and process improvement. Several 2.5D and 3D SiP-packaging platforms have emerged and appear to be ready for production.

References

1. Priest, J., M. Ahmad, L. Li, J. Xue, and M. Brillhart, Design optimization of a high performance FCAMP package for manufacturing and reliability, *Proceedings 55th Electronic Components and Technology Conference*, Orlando, FL, May 2005, pp. 1497–1501.
2. Priest, J. and L. Li, Challenges in substrate design, assembly, and reliability of SiP package for a high end networking application, *39th International Symposium on Microelectronics*, IMAPS, San Diego, CA, October 2006, pp. 8–12.
3. Li, L., S. Peng, J. Xue et al., Addressing bandwidth challenges in next generation high performance network systems with 3D IC integration, *Proceedings of the 62nd Electronic Components and Technology Conference (ECTC)*, San Diego, CA, May 2012, pp. 1040–1046.
4. S. Lakka, Xilinx SSI technology, *Hot Chips: A Symposium on High Performance Chips*, Palo Alto, CA, August 2012.
5. Li, L., P. Chia, P. Ton, et al., 3D SiP with organic interposer for ASIC and memory integration, *Proceedings of the 66th Electronic Components and Technology Conference (ECTC)*, Las Vegas, NV, June 2016, pp. 1445–1450.
6. JEDEC Standard, Addendum No. 3 to JESD79-3: 3D Stacked SDRAM, JESD79-3-3, December 2013.
7. JEDEC Standard, Wide I/O single data rate (wide I/O SDR), JESD229, December 2011.
8. JEDEC Standard, Wide I/O 2, JESD229-2, August 2014.
9. Hybrid Memory Cube. Retrieved from http://hybridmemorycube.org/specification-v2-download-form/

10. JEDEC Standard, High bandwidth memory (HBM) DRAM, JESD235A, November 2015.
11. TSMC. CoWoS® Services. Retrieved from http://www.tsmc.com/english/ dedicatedFoundry/services/cowos.htm
12. Yoon, S. W., D. J. Na, W. K. Choi et al., 2.5D/3D TSV processes development and assembly/packaging technology, *Proceedings of the 11th Electronics Packaging Technology Conference (EPTC)*, Singapore, December 2011, pp. 336–340.
13. Sunohara, M., A. Shiraishi, Y. Taguchi, K. Murayama, M. Higashi, and M. Shimizu, Development of silicon module with TSVs and global wiring (L/S = 0.8/0.8 μm), *Proceedings of the 59th Electronic Components and Technology Conference (ECTC)*, San Diego, CA, May 2009, pp. 25–31.
14. Li, L., M. Higashi, A. Takano, J. Xue, and G. Ikari, Cost and performance effective silicon interposer and vertical interconnect for 3D ASIC and memory integration, *Proceedings of the 64th Electronic Components and Technology Conference (ECTC)*, Orlando, FL, May 2014, pp. 1366–1371.
15. JEDEC Product Outline, MO-316, HBM micropillar grid array package (MPGA), October 2015.
16. JEDEC Publication, JEP158 3D chip stack with trough-silicon vias (TSVS): Identifying, evaluating and understanding reliability interactions, November 2009.
17. JEDEC Standard, JESD47, Stress-test-driven qualification of integrated circuits, 2015.
18. Li, L., P. Ton, M. Nagar, and P. Chia, Reliability challenges in 2.5D and 3D IC integration, *Proceedings of the 67th Electronic Components and Technology Conference (ECTC)*, Orlando, FL, May 2017, pp. 1504–1509.

3

A New Class of High-Capacity, Resource-Rich Field-Programmable Gate Arrays Enabled by Three-Dimensional Integration Chip-Stacked Silicon Interconnect Technology

Suresh Ramalingam, Henley Liu, Myongseob Kim,
Boon Ang, Woon-Seong Kwon, Tom Lee, Susan Wu,
Jonathan Chang, Ephrem Wu, Xin Wu, and Liam Madden

CONTENTS

3.1 Introduction

As the role of the field-programmable gate arrays (FPGA) becomes more significant in larger and more complex system designs, it demands higher logic capacity and more on-chip resources and functionalities. Until 40 nm node, FPGAs have depended predominantly on Moore's law scaling to respond to this need, delivering nearly twice the logic capacity with each new process generation. However, keeping pace with today's high-end market demands requires more than Moore's law can provide.

To respond to these requirements, Xilinx first introduced a three-dimensional integration chip (3D-IC) stacked silicon interconnect technology (SSIT) for building FPGAs that offer bandwidth and capacity that are not realized by traditional Moore's law scaling (Figure 3.1). SSIT uses silicon interposers with microbumps (μBumps) and through-silicon vias (TSV) to integrate multiple FPGA die slices in a single package.

Starting from 2006 (Figure 3.2), Xilinx worked with research consortiums, equipment vendors, foundry, and outsourced assembly and test (OSAT) partners to develop TSVs, μBump, bonding, and 3D-IC integration technology.

Xilinx's first 3D-IC SSIT FPGA, Virtex®-7 2000T, introduced in late 2011 mainly for emulation applications has been well adopted by the industry. 3D-IC devices since have expanded into networking, data center cloud computing, and high-performance computing (HPC) areas in volume production and account for more than 30% of the 20 nm/16 nm products offered. The largest SSIT FPGA as of today is UltraScale XCVU440 with 4.4 million logic cells, 600 k μBumps, 19 billion transistors in a 55 mm package, and has been shipping in volume production since early 2016.

This chapter describes Xilinx's 3D-IC development experiences gained by launching process learning vehicles, developing design simulation methodology, pursuing early-stage reliability assessment, and optimizing

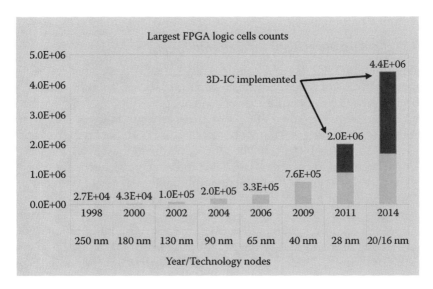

FIGURE 3.1
FPGA capacity doubling as 28 nm generation enabled by 3D-IC.

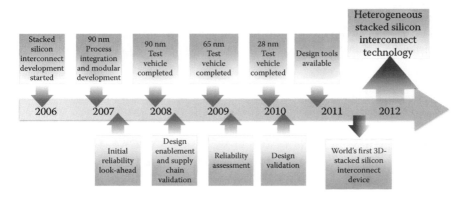

FIGURE 3.2
FPGA 3D-IC history—from concept to world's first programmable products on 28 nm.

the process baseline. All the development efforts paved the foundation for a robust production line. At the end, the potential future 3D-IC technology challenges are outlined.

3.2 Architecture, Design, and Product Enablement

Let us start with a summary of the challenges faced implementing large designs with multiple monolithic FPGAs and how SSIT addresses them [1].

3.2.1 Key Challenges

3.2.1.1 Limited Connectivity and Bandwidth

System-on-chip (SoC) designs comprise millions of logic gates connected by complex networks of wires in the form of multiple buses, complicated clock distribution networks, and multitudes of control signals. Successfully partitioning a SoC design across multiple FPGAs requires an abundance of I/Os to implement the nets spanning the gap between FPGAs. With SoC designs including buses as wide as 1024 bits, even when targeting the highest available pin-count FPGA packages, engineers must use data buffering and other design optimizations that are less efficient for implementing the thousands of one-to-one connections needed for high-performance buses and other critical paths.

Packaging technology is one of the key factors to this I/O limitation. The most advanced packages currently offer approximately 1200 ~ 1500 I/O pins, far short of the total number of I/Os required.

At the die level, I/O technology presents another limitation because I/O resources do not scale at the same pace as interconnect logic resources with each new process node. When compared to transistors used to build the programmable logic resources in the heart of the FPGA, the I/O transistors are much larger to deliver the currents and withstand the voltages required for chip-to-chip I/O standards. Thus, increasing large number of standard I/Os on a die is not a viable solution to provide the connections for combining multiple FPGA die.

3.2.1.2 Excessive Latency

Increased latency is another challenge with the multiple FPGA approach. Standard device I/Os impose pin-to-pin delays that degrade the overall circuit performance for designs that span multiple FPGAs. Moreover, using time-domain multiplexing (TDM) on standard I/Os to increase the virtual pin count by running multiple signals on each I/O imposes even greater latencies that can slow I/O speeds further down by a factor of 4x–32x or more.

3.2.1.3 Power Penalty

Connecting multiple FPGAs via standard I/Os approaches also results in higher power consumption. When used to drive hundreds of package-to-package connections across printed circuit board (PCB) traces, standard I/O pins carry a heavy-power penalty compared to connecting logic nets on a monolithic die.

Similarly, multichip module (MCM) technology offers potential form-factor reduction benefits for integrating multiple FPGA die in a single package. The MCM approach, however, still suffers from the similar restrictions of limited I/O count and undesirable latency and power consumption characteristics.

3.2.2 Xilinx-Stacked Silicon Interconnect Technology

To overcome these limitations and roadblocks, Xilinx developed a new approach that enables high-bandwidth connectivity between multiple dies by providing a much greater number of interdie connections. It also imposes much lower latency and consumes dramatically lower power than the multiple FPGA via standard I/Os approach, while enabling the integration of massive quantities of interconnect logic and on-chip resources within a single package.

Xilinx arrived at such a solution by applying several proven technologies in an innovative way. By combining TSV and µBumps with its application-specific modular block (ASMBL™) architecture, Xilinx is building a new class of FPGAs that delivers the capacity, performance, capabilities, and power characteristics required to address the programmable imperative. Figure 3.3 shows the top view of the die stack-up with four FPGA die slices, silicon interposer, and package substrate. SSIT technology combines enhanced FPGA die slices and a passive silicon interposer creates a die stack that implements tens of thousands of die-to-die connections to provide ultrahigh interdie interconnect bandwidth with far lower power consumption and one fifth of the latency of standard I/Os.

Being originally developed for use in a variety of die-stacking design methodologies, silicon interposers provide modular design flexibility and high-performance integration suitable for a wide range of applications. The silicon interposer acts as a sort of microcircuit board in silicon on which multiple dies are set side by side and interconnected. Compared to organic or ceramic substrates, silicon interposers offer far finer interconnect geometries (approximately 20X denser wire pitch) to provide device-scale interconnect hierarchy that enables more than 10,000 die-to-die connections.

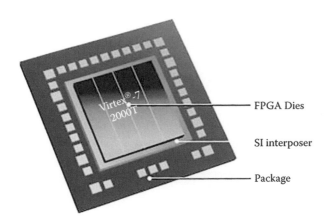

FIGURE 3.3
Stacked silicon interconnect package top view.

3.2.2.1 Creating Field-Programmable Gate Array Die Slices with Microbumps for Stacked Silicon Integration

The foundation of Xilinx SSIT technology is the company's ASMBL architecture, a modular structure comprising Xilinx FPGA building blocks in the form of tiles that implement key functionality such as configurable logic blocks (CLBs), block RAM, digital signal processing (DSP) slices, SelectIO™ interfaces, and serial transceivers. Each type of tile is organized into blocks in columns and then combines the columns to create an FPGA. By varying the height and arrangement of columns, assortment of FPGAs can be created with different amounts and mixes of logic, memory, DSP, and I/O resources (Figure 3.4). The FPGA contains additional blocks for generating clock signals and for programming the static random access memory (SRAM) cells with the bitstream data that configure the device to implement the end user's desired functionality. The architecture can be extended for heterogeneous die integration as well [2,3].

Starting with the basic ASMBL architectural construct, three key modifications that enable stacked silicon integration (Figure 3.5) have been introduced. First, each die slice receives its own clocking and configuration circuitry. Then the routing architecture is modified to enable direct connections to routing resources within the FPGA's logic array, bypassing the traditional parallel and serial I/O circuits. Finally, each slice undergoes additional processing steps to fabricate microbumps that attach the die to the silicon interposer. All of these enable connections in far greater

FIGURE 3.4
FPGA tiles built with ASMBL™ architecture.

FIGURE 3.5
FPGA die or SLR (super logic region) optimized for stacked silicon integration.

numbers, with much lower latency, and much less-power consumption than is possible using traditional I/Os (100x the die-to-die connectivity bandwidth per watt versus standard I/Os).

3.2.2.2 Silicon Interposer with Through-Silicon Vias

The passive silicon interposer interconnects the FPGA die. It is built on a low-risk, high-yielding 65 nm process and provides four layers of metallization (3 Cu and 1 Al) for building the tens of thousands of traces that connect the logic regions of multiple FPGA die (Figure 3.6).

Figure 3.7 illustrates the concept of the assembled die stack. It contains a stack-up of four FPGA die mounted side by side on a passive silicon interposer

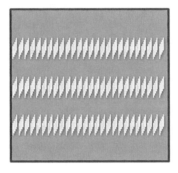

FIGURE 3.6
Passive silicon interposer with die-to-die interconnect wiring.

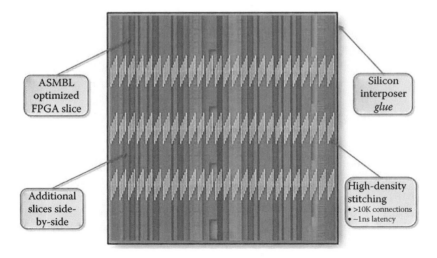

FIGURE 3.7
Transparent view of the assembled die stack and die-to-die interconnects using silicon interposer.

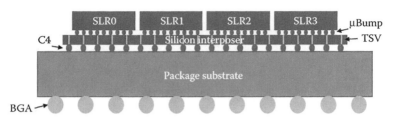

FIGURE 3.8
Cross-section view of FPGA-silicon interposer integration on a package substrate.

(bottom view). The interposer is shown as transparent to enable a view of the FPGA die slices connected by traces on the silicon interposer (not to scale).

The TSVs combined with controlled-collapse chip connection (C4) solder bumps enable mounting the FPGA/interposer stack-up on a high-performance package substrate using flip-chip assembly techniques (Figure 3.8). The coarse-pitch TSVs provide the connections between the package and the FPGA for the parallel and serial I/O, power/ground, clocking, configuration signals, and so on.

3.2.2.3 Three-Dimensional Integration Chip Analysis Methodology Enablement with Simulation Program with Integrated Circuit Emphasis Simulators

As with any other new technology, 3D-IC also brings new challenges to traditional ad hoc design simulation tools because of the integration of multiple dice and technologies. In order to take full advantage of 3D-IC integration, Xilinx initiated and codeveloped with Synopsys a new simulation program with integrated circuit emphasis (SPICE) simulation methodology for 3D-IC module in 2012.

Although traditional circuit simulation tools work well on single-die simulations, it is challenging to handle multidie simulations. One challenge is in the common naming convention, such as global node names, subcircuit (subckt) names, and model names. Global parameters confliction brings another challenge. Extra effort is needed to modify the netlists and technology models, making it a very time-consuming and error-prone process. In addition, this workaround cannot handle different technologies with varied process parameters such as tnom, scale, and geoshrink not to mention for third-part IPs where the netlists are encrypted, it becomes impossible to do any netlist modifications.

In contrast, the new 3D-IC features taking on a bottom-up approach provide a complete integrated simulation solution that works for any combination of integrated technologies.

By introducing module block syntax for each die, the approach enables multidice chip-level simulations under the existing SPICE simulation environment. Besides providing the simulation capabilities for 3D-IC designs,

```
xic1 n1 n2 n3 mod1::top_sub
xic2 n1 n2 n3 n4 mod2::top_sub

/* Multi-die, multi-technology and multiple thermo domain */
.module mod1
    .temp temp1    /* operating temperature in IC module 1 */
    .model pmos pmos ...   /* 20nm MOS model */
        .subckt top_sub p1 p2 p3
            m1 p1 p2 p3 p4 pmos ...   /* 20nm technology MOS */
            ...
        .ends
    ...
.endmodule mod1

.module mod2
    .temp temp2    /* operation temperature in IC module 2 */
    .model pmos pmos ...   /* 40nm MOS model */
        .subckt top_sub p1 p2 p3 p4
            X1 p1 p2 p3 p4 inv ...   /* 40nm technology INV */
            ...
        .ends
    ...
.endmodule mod2
```

FIGURE 3.9
Module-based netlist for 3D-IC integration chip with two dice from different technologies.

the major advantage of the new approach over the traditional one is its capability to maintain libraries and netlist files to help reduce memory usage and simulation errors.

An example netlist in Figure 3.9 below illustrates the simple 3D IC integration chip with two dice of different technologies and temperatures as "xic1" and "xic2."

Each design team is expected to exercise the full design verification flow for each IC module before the 3D-IC integration. As each integrated IC module is regarded as a completed product, retaining the IC module netlist integrity becomes a primary requirement. New HSPICE netlist features are then introduced to target the multiple-die and multitechnology integration for both heterogeneous and homogeneous designs [4].

New 3D-IC-specific configurations and analysis requirements are introduced as well, such as multidie multitechnology integration, verification with multiple operational IC temperature domain, and exponentially increasing corner simulations for multidie integration with different simulation corners on each IC module.

For both prelayout and postlayout design analysis and verification, the new SPICE simulators offer 3D-IC-specific methodology for functional verification, timing verification, and power analysis.

The flow chart in Figure 3.10 provides the steps for 3D-IC integration simulations using Synopsys' 3D-IC simulation features. The new 3D-IC simulation features and the methodology offer an efficient simulation solution for timing closure of multidie and multitechnology integration under the conventional SPICE simulation environment.

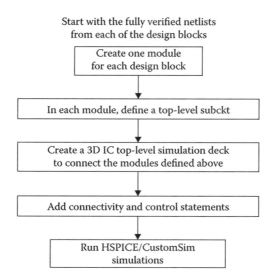

Start with the fully verified netlists
from each of the design blocks

Create one module
for each design block

In each module, define a top-level subckt

Create a 3D IC top-level simulation deck
to connect the modules defined above

Add connectivity and control statements

Run HSPICE/CustomSim
simulations

FIGURE 3.10
3D-IC SPICE simulation flow.

Silicon interposer signal integrity also requires careful consideration both from transmission and signal loss perspective and there are a few Xilinx publications on this topic [2,5,6]. In Section 3.2.2.4, we will examine homogeneous and heterogeneous signal integrity analysis partitioning used successfully by Xilinx to deliver 4 × 100 GB/s line card solution. It is to be noted that this case study used a low-temperature cofired ceramic (LTCC) substrate though actual product implementation was done using a low-loss organic substrate.

3.2.2.4 Silicon Interposer Signal Integrity

Although the silicon interposer is only hundred microns thick, failure to consider high-frequency effects of the TSVs will degrade the rise/fall time, increase cross-talk, increase noise injection, and cause significant performance degradation of the signal transmitted through the high-speed channel. In addition, die-to-die signals that pass through the microbumps and laterally through the fine-pitch metallization also need careful consideration.

This section summarizes signal integrity studies for two types of off-die signals in 3D designs: (1) Type I and (2) Type II signals. A Type I signal connects one or more active ICs and a package pin. A Type II signal supports interdie communication only and is not connected to any package pins. For clarity, Type I signals are electrically connected to TSVs in the silicon interposer, whereas Type II signals are not.

FIGURE 3.11
Signal Type I (vertical to package pin, as "a") and Type II (interdie, as "b" and "c").

Signal "a" in Figure 3.11 is a Type I signal. Typically, multiple microbumps are connected to one C4 bump to ensure adequate electromigration margin. Signal "b" and "c" are Type II signals from one top die to another top die.

Sections 3.2.2.4.1 and 3.2.2.4.2 discuss signal integrity related to the GTZ where GTZ is a stand-alone 28 Gbps SerDes die and super logic regions (SLRs) are 2 FPGA dies as described previously.

3.2.2.4.1 Type I Signals

The GTZ-IC receives eight differential serial streams at up to 28.05 GB/s each, deserializes them with serial-in-parallel-out (SIPO) blocks, and sends the parallel data and necessary clocks to the SLR. Similarly, when in output, it receives parallel data from the SLR, serializes the traffic with parallel-in-serial-out (PISO) blocks, and outputs the resulting differential serial streams.

The GTZ Type I signals are in the analog supply domain, which has its own power and ground metal planes. The parallel Type II signals are in the digital supply domain, sharing the digital and ground planes with the SLR. Analog power and ground decoupling capacitors are distributed on the die, on the interposer, and on the package substrate to ensure optimal signal integrity at 28 GB/s.

3.2.2.4.1.1 Silicon Interposer TSV Modeling and Optimization A full 3D electromagnetic (EM) field solver is used to model the silicon interposer TSV accurately over a wide frequency range. A broadband S-parameter model is generated with an upper frequency limit of 50 GHz. The interposer consists of TSVs, four metal layers for die-to-die connections, microbumps for die-to-interposer connections and C4 bumps for interposer-to-package connections. All components are incorporated into the model to predict actual interposer performance in the system. A silicon interposer test vehicle was fabricated for measurement and verification (Figure 3.12). Various TSV test structures were characterized across a wide frequency range. The TSV is approximately 100 μm in height. De-embedding microprobe effects is critical for accurate frequency-domain vector network analyzer (VNA) measurements. As a first step, a microprobe calibration is performed to move the reference plane to the end of microprobe. Depending on the metallization of the probe tip and the surface of the pad, a residual, uncompensated resistive impedance in the order of 10 ~ 100 $m\Omega$ could still be present. The flexing or overtravel of the

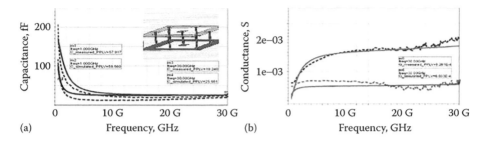

FIGURE 3.12
Simulation versus measurement comparison for (a) effective capacitance and (b) effective conductance.

microprobe can also result in an uncompensated, residual inductance of as much as 100 pH. To overcome these parasitic, a 2-port VNA was used.

The 2-port VNA eliminates the residual contact resistance and inductance. However, mutual magnetic coupling between probes still exists if probes are located face to face in close proximity. To improve high-frequency measurement accuracy, 90° orthogonal probing is performed to cancel the magnetic field coupling from the two probes. These result in improved measurement especially for higher frequencies (>5 GHz).

As the focus is on multigigabit operation within the interposer, the model must support and have a good agreement for frequencies up to tens of gigahertz. The measured data are compared with simulation results. Figure 3.12 shows the comparative results of effective capacitances which were extracted from the open test structures. Two different silicon substrates are fabricated to compare the substrate resistivity effects. One test interposer sample has a 10 Ω-cm silicon resistivity, representative of a typical commercial process and the other test interposer sample has a 20 Ω-cm silicon resistivity.

In Figure 3.12, the dotted traces represent the measured data and the solid traces represent the simulated data of effective capacitance and conductance values over frequency from the 10 Ω-cm and 20 Ω-cm silicon resistivity substrates, respectively. Good agreement is observed between the measured and simulated results. The effective capacitance decreases as a function of frequency (slow-wave model to quasi-TEM mode). This effect can be modeled as resistance and capacitance in the silicon. The measured effective capacitance from the 10 Ω-cm silicon resistivity substrate was about 205.2 fF at low frequency and 19.4 fF at high frequency, whereas from the 20 Ω-cm silicon resistivity substrate, the measured effective capacitance was about 115.8 fF at low frequency and 12.4 fF at high frequency. Both low- and high-frequency capacitance are important electrical parameters because they directly affect data eye opening, power consumption, and delay.

A second through test structure is fabricated to evaluate insertion for serially connected TSVs. S21 insertion loss is measured and compared with simulation as shown in Figure 3.13. The measured insertion loss at the Nyquist frequency

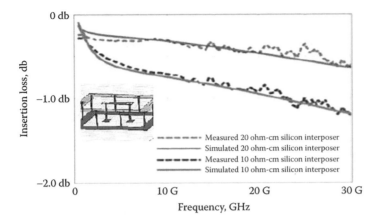

FIGURE 3.13
Simulation versus measurement comparison for insertion loss.

(14 GHz) is −0.414 dB from the 20 Ω-cm silicon substrate and −0.822 dB from 10 Ω-cm silicon substrate, respectively. The high-resistivity silicon substrate obviously provides lower loss over the entire frequency range.

Based on these measurements and simulation results, we conclude that a 20 Ω-cm high-resistivity silicon substrate is preferred for very high-speed signaling applications. Higher than 20 Ω-cm resistivity silicon substrates are also available in the industry if there is a need to further reduce TSV insertion loss.

3.2.2.4.2 Type II Signals

3.2.2.4.2.1 Silicon Interposer Signal Routing   The silicon interposer lateral routing consists of four layers of metal: one layer for redistribution, one for ground reference, and two for signal routing. The ground reference layer separates the two signal-routing layers. In each signal-routing layer, every signal wire runs parallel to a grounded side shield.

Figure 3.14 illustrates the effect of these grounded side shields on a victim wire. Eleven interposer wires of 4.5 mm in a single metal layer on top of a ground reference layer are simulated. The victim wire is at the center of the eleven-wire bus. The interposer wires are modeled with 3D RLC elements. The simulation consists of 10 aggressors transitioning with a range of delay values relative to the victim. Figure 3.14a and b illustrates the victim waveform without and with side shields inserted into the eleven-wire bus to eliminate overshoots and undershoots, respectively.

3.2.2.4.2.2 Interdie Timing Verification   Interdie routing requires rigorous timing analysis to verify adequate setup and hold time.

There are over 4000 Type II signals between the GTZ and its neighboring SLR. Figure 3.15 illustrates the wire length distribution of these Type II signals. All but five signals are routed on traces no more than 6 mm in length,

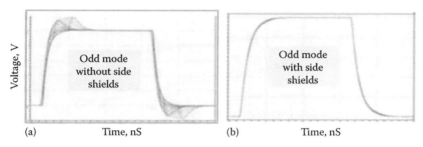

FIGURE 3.14
Even-mode coupling (a) Even mode without side shields and (b) Even mode with side shields.

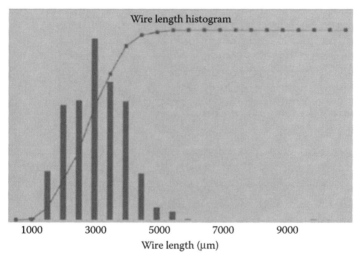

FIGURE 3.15
Type II signal wire length distribution.

whereas 85% of the traces are less than 3.75 mm in length. Clocking between the GTZ and its neighboring SLR is system synchronous. Careful balancing of clock networks on these two pieces of silicon across process, voltage, and temperature is necessary. Static timing analysis (STA) verifies interdie timing efficiently. However, during XC7VH580T development, STA tools did not accept RLC interconnect models. As a result, interdie propagation delay and transition time values were calibrated between SPICE simulations with 3D RLC interconnect models and STA with RC interconnect models. The layout ensures that Type II signaling in the XC7VH580T is not in transmission-line mode and STA with RC-based interconnects suffices.

3.2.3 Three-Dimensional Integration Chip Resource-Rich Field-Programmable Gate Arrays Product Offerings

SSIT provides multi-terabit-per-second die-to-die bandwidth through more than 10,000 device-scale connections—enough for the most complex multidie designs. Xilinx first used this technology to create Virtex®-7 (28 nm) FPGA products in 2011, which offered unprecedented capabilities including up to 2 million logic cells; 65 Mb of block RAM; 2375 GMACS of DSP performance (4750 GMACS for symmetric filters); 1200 SelectIO pins supporting 1.6 GB/s low-voltage differential signaling (LVDS) parallel interfaces; and 72 serial transceivers delivering 1886 GB/s aggregate bidirectional bandwidth.

Besides homogeneous partitioning, Xilinx also qualified and productized heterogeneous 3D integration via silicon interposer of third-party 28 Gbps SerDes dies in 28 nm generation [2,7]. XC7VH580T and XC7VH870T are two heterogeneous 3D-IC integration products introduced in 2013. Other examples include internally designed analog-to-digital converter/digital-to-analog converter (DAC/ADC) heterogeneously integrated on an interposer with 2 FPGA SLRs, which was demonstrated at ISSCC 2014 [3].

Since then, Xilinx has expanded the 3D-IC technology envelope further with Ultrascale (20 nm) and Ultrascale+ (16 nm fin field-effect transistor [FINFET]) products to support emulation, 200G/400G and datacenter applications. New development such as larger-than-reticle interposer (X-interposer) with lithography-stitched metal lines and 600,000 µBumps interconnections [8–10] have been qualified and are in volume production. Homogeneously partitioned lithography-stitched devices in production include XCVU440, XCVU190, XCVU11P, and XCVU13P. High bandwidth memory (HBM)-integrated products have also been announced [8].

3.3 Stacked Silicon Technology Development and Package Reliability

To meet the high 3D-IC FPGA demand, the process line in the fab needs to maintain high yield and reliability all the time. 3D-IC FPGA assembly yield improvement was discussed by B. Banijamali et al. [7,9], R. Chaware et al. [11] and reliability was discussed by L. Lin et al. [12], respectively. New process modules are also needed to enable silicon interposer TSV reveal, the so-called middle-end-of-line (MEOL) steps. Section 3.3.1 describes key-enabling technologies and how a well-defined test vehicle played a significant role in Xilinx's 3D-IC technology development and reliability evaluation.

3.3.1 Key-Enabling Technologies

One possible silicon interposer MEOL scheme (soft reveal) is illustrated in Figure 3.16. Other approaches such as flat reveal are also commonly employed. The main difference is whether the TSV is protected by a capping layer or not during silicon grinding. Nevertheless, these various process schemes have several common process steps—microbump, carrier mount,

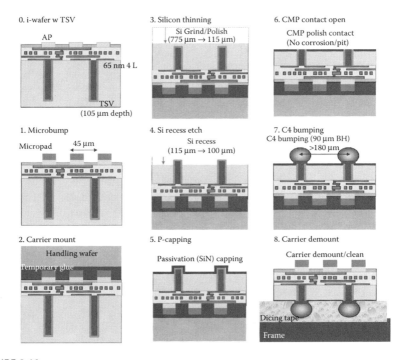

FIGURE 3.16
MEOL soft reveal process flow scheme: 0—i-wafer with TSV, 1—Microbump, 2—Carrier mount, 3—Silicon thinning, 4—Si recess etch, 5—P-capping, 6—CMP contact open, 7—C4 bumping, and 8—Carrier demount.

silicon thinning, Si etch, backside passivation, chemical–mechanical polishing (CMP) contact open, C4 bumping, and carrier demount.

The μBump formation requires significant interface optimization for tight resistance control. As the critical dimensions (CD) are much smaller than regular C4 Cu-pillar bumps (for e.g., 25 μm vs. 50 μm), the sensitivity of resistance and reliability to process variations and defect is higher.

Many factors can cause considerable yield loss in foundry TSV manufacturing and OSAT MEOL processes, including the different definitions of critical killer defects from existing fab/OSAT integration, glue residue/contamination at bevel and edge areas due to plasma inhomogeneity at wafer edge, wafer bow due to film stress, nonuniformity in etch profiles and post-CMP film thickness, and plasma-induced or handling-induced damage at thin wafer edge. Considerable efforts have been made to enhance line yield and interposer die yield in various aspects of particle control, wafer-edge engineering, process uniformity improvement, surface treatment/modification, thin wafer handling damage, and so on. From a process integration perspective, statistical process control (SPC), which is extremely useful in process control and optimization, is also applied for yield improvement in MEOL module. The following are several case studies illustrating yield improvement efforts and substantial enhancement outcome [13].

3.3.1.1 Silicon-Grinding Quality Optimization

During the MEOL process, the front side of the TSV interposer wafer is temporarily glued to a carrier wafer. Bonding glue total thickness variation (TTV), edge bead amount, glue coverage for trimmed bevel, and bonding void are optimized to improve line yield and accommodate subsequent process integration. TSV interposer wafers are ground with successively finer grade abrasives (Z1/Z2/Z3) to thin down until a certain amount of silicon is left on TSV copper bottom. Post-grind surface property of silicon is very critical for particle control during subsequent process steps. For example, hydrophobic Si surface tends to repel water droplets that leave organic material. When the water droplet is removed from the hydrophobic surface during wafer-drying process, the watermarks remained and organic residues are formed. This watermark defect is invisible right after grinding step and appears after reactive ion etching (RIE) step. This is a very crucial defect mode for subsequent RIE etch-back and final known good die (KGD) electrical testing. It is necessary to remove the watermark and contaminants from the post-grind Si surface. Figure 3.17 shows that both Si CMP optimization and post-grind surface clean could result in defect-free surface. The control of the wettability of post-grind surface and post-CMP cleaning method plays an important role on the removal of particles and watermarks from Si surface.

3.3.1.2 Wafer Edge and Bevel-Cleaning Optimization

During the MEOL manufacturing process, defects, carrier wafer damages, and contamination are likely to occur at wafer edge caused by temporary glue particulate, dry etching, film deposition, and wafer handling. These edge defects and particles are easily transferred to polished silicon surface during subsequent process steps, for example, RIE, chemical vapor deposition (CVD), or wafer transport steps. Figure 3.18 depicts the representative defect images after RIE etch-back step. Foreign material (a) could come from

Method	CMP 1 +CLN1	CMP 2 +CLN1	CMP 2 +CLN2 (POR)
Surface contact angle	Hydrophobic	Hydrophilic	Hydrophilic
Defect map	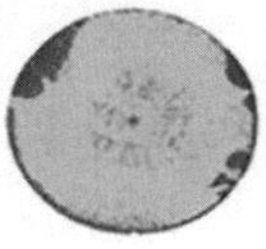		
Defects	Watermark, chemical residue	Watermark	No defect (100% yield)

FIGURE 3.17
Defect improvement before and after postgrind surface optimization.

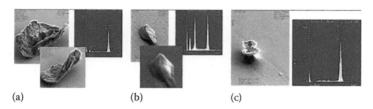

(a)   (b)   (c)

FIGURE 3.18
Defect types (a) foreign material, (b) Sulfur hexafluoride (SF_6) by-product, and (c) pre-existing particle masking defect.

either edge glue particulate or environment random defects. Etch by-product (b) is the in-process particle. Hard mask defect (c) can originate from pre-etch defects that act as Si hard mask. Wafer edge or bevel area is a primary source of yield lowering defects. Therefore the cleanliness at wafer edge (top and bevel) is crucial for yield enhancement.

A special wafer edge bevel cleaning (wet + dry) technique can be effective in reducing defect source from wafer edge during RIE & CVD steps. In order to see the impact of bevel clean in single RIE etch step, different bevel cleaning methods prior to RIE etch-back step were evaluated. Figure 3.19 shows defect reduction and yield improvement due to bevel clean. The minor fall-on particles shown in Figure 3.19c could be scrubbed away in subsequent wafer cleaning step.

3.3.1.3 Silicon-Footing Improvement

Plasma etch has localized etch rate difference between dissimilar materials (Si and SiO_2) interface due to etch radical charging on oxide surface that causes either Si footing or notching. It becomes much worse with over-recessed TSV tip height (or pillar height), which is more likely to occur at wafer edge. Figure 3.20a displays the worst-case electrical failure case due to Si footing defect particularly at wafer edge.

In order to enhance TSV tip height uniformity after RIE step, gradient Si etch rate across the wafer is applied to compensate for the thickness

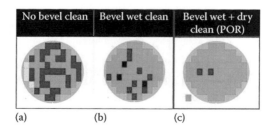

(a)   (b)   (c)

FIGURE 3.19
Die yield improvement comparing (a) without bevel clean, (b) bevel wet clean, and (c) bevel wet + dry clean.

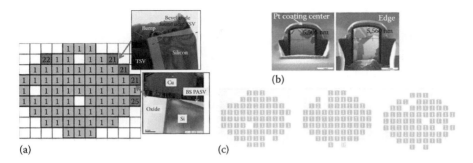

FIGURE 3.20
(a) Interposer wafer edge failures caused by nonoptimized etching revealed, (b) center and edge revealed TSVs after Si gradient etch implementation, and (c) no failures after etch optimization.

nonuniformity on incoming wafers due to carrier bond and lapping steps. The focused ion beam (FIB)/SEM images in Figure 3.20b show revealed TSV tip with optimized RIE etch-back. It proves that TSV tip height variation across wafer (center to edge) is uniform and oxide liner loss is little on top of exposed TSV tip. With aforementioned gradient Si etching technique and localized charge reduction effort, the TSV tip height uniformity and silicon footing size are substantially improved and final KGD yield is almost closed to ~100% as shown in Figure 3.20c.

3.3.2 Three-Dimensional Integration Chip Development Test Vehicles

Test chips or test vehicles (TVs) are important to understand technology, design, and/or their interactions before a new product launch. 3D-IC is an integration of new silicon, packaging, and 3D-IC technologies together, thus test chips play especially an important role. Xilinx works closely with foundry by running multiple test chips in almost all technologies (Figure 3.21).

Year	20XX												20XX+1											
Quarter	Q1			Q2			Q3			Q4			Q1			Q2			Q3			Q4		
Month	1	2	3	4	5	6	7	8	9	10	11	12	1	2	3	4	5	6	7	8	9	10	11	12
SPICE	V0.01						V0.1						V0.5						V0.5					
DRM/DRC	V0.01						V0.1						V0.5						V0.5					
TV (MPW)	TV1 TO ☆						TV2 TO ☆						TV3/3D-IC TO ☆											

FIGURE 3.21
Milestone of PDK delivery and test vehicle development for new technology node and 3D-IC.

3.3.2.1 28 nm Test Vehicle-Driven Process and Reliability Improvements

This section discusses the 28 nm 3D-IC TV. The top die of the 28 nm 3D-IC test vehicle (TV) was populated by 1.8 V I/O blocks with long passive scan chains. Each of these long passive scan chains in a block has ~2.7 K of microbumps. Each test chip unit contains ~110 K microbumps. Each 3D-IC test chip consists of four top dies on an interposer (Figure 3.22).

Constrained by resources in this TV, the top die could not be probed at wafer level to check the yield. The yield was validated at chip level and found it was relatively straightforward to differentiate the yield scan chain versus open/short scan chain electrically. However, the physical failure analysis (FA) feedback loop takes longer time to identify the failure site from the long chain. The typical 3D-IC FA flow used in Xilinx is illustrated (Figure 3.23).

After e-test at chip level, x-ray and confocal scanning acoustic microscopy (C-SAM) detection are the frequently used constructive inspection approaches to validate the test results, in particular failing or delamination locations. In Figure 3.24, one of the typical FA case results is shown.

Detecting microbump interconnect robustness is a main focus in Xilinx's 28 nm 3D-IC TV as the high microbump count (~200,000 in largest device) requires joining yield to be extremely high. The TV helped to detect either microbump process systematic (defect, particle/residue, chip warpage) or excursion issues during the development period. If looking back and weighing the significance of the 3D-IC TV, one would recognize it as a very successful 3D-IC TV that paved the foundation of microbump process baseline.

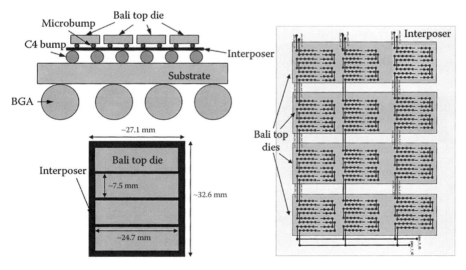

FIGURE 3.22
A schematic of Xilinx 28 nm 3D-IC TV (code named "Bali").

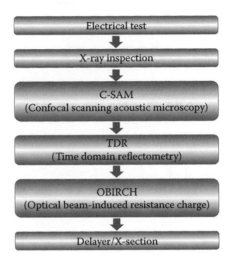

FIGURE 3.23
A typical 3D-IC physical FA working flow.

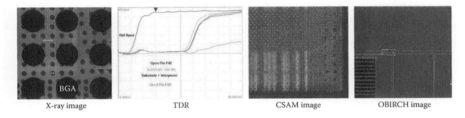

FIGURE 3.24
One typical FA case using X-ray and C-SAM inspection approaches.

In fact the TV helped with choosing chip-on-wafer (CoW) as the preferred assembly integration platform (Figure 3.25). In Figure 3.26, typical microbump open/short cases are demonstrated.

3.3.2.2 Improvements to 20 nm Test Vehicle

In Xilinx's 20 nm 3D-IC TV, to meet the product features (microbump scaling from pitch 45–40 μm with 600,000 microbumps and X-interposer that has a size larger than a regular lithography reticle size), long passive scan chain was replaced by defect-monitoring vehicle (DMV). Xilinx has been using DMV for many generations due to its rapid defect detection capability and accuracy. DMV is large and sensitive enough to catch PPM-level open and short defects. The DMV was redesigned to fit into 3D-IC development for X-interposer with 40 μm pitch microbump defect detection. Figure 3.27 illustrates the floor plan of X-interposer with 3 slices of DMV (each slice is 23.25 × 14.56 mm) that contains a total of ~375 K microbumps.

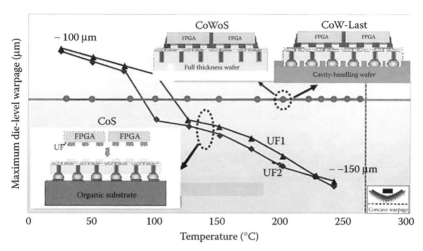

FIGURE 3.25
Warpage behavior during microbump joining for different integration schemes.

FIGURE 3.26
Demonstration of one typical microbump open and short FA case, respectively.

As expected, DMV's failing location identification capability and efficiency assisted to identify open/short microbump defect. In particular, the DMV can have wafer-level test capability implemented so the KGD can be achieved before 3D-IC integration on X-interposer. This 20 nm 3D-IC TV also helped to identify the μBump density variation versus yield sensitivity, which led to a specific design design for manufacturability (DFM) establishment.

Board-level reliability (BLR) of package is another important aspect to consider from a system-use perspective. Introducing new technology elements such as μBump, multidie, X-interposers, and large package sizes requires a careful study of failure modes to ensure a good understanding of product performance in cycling, shock, bend, and other tests.

Xilinx, working with our lead customer inputs, defined a special vehicle with hooks that go well beyond traditional ball grid array (BGA) daisy-chain monitoring to include C4, TSV, μBumps, and die edge crack monitoring. The focus was on indoor Telecom use conditions that are stringent due to use of

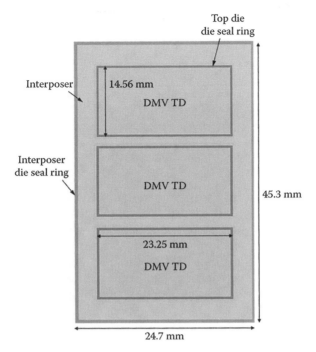

FIGURE 3.27
20 nm 3D-IC TV with 3 DMV top dies on X-interposer containing ~375 K μBumps.

very thick, high-layer count boards (high stress on BGA), and long life time expectation. In Section 3.3.2.3, we briefly discuss the vehicle and results seen in board level 0°C–100°C thermal cycling and Shock and Bend.

3.3.2.3 Board-Level Reliability Test Vehicle and Results

A specially designed test vehicle was developed to conduct BLR test. Traditionally, per IPC-9710, BGA solder balls are monitored through daisy chains in both package peripheral and under die perimeter areas. For the X-interposer package, daisy chains were added to monitor C4 bumps, TSV, and microbumps. To achieve this purpose, top die was redesigned, interposer and substrate from the original component package, as shown in Figure 3.28, which illustrates a daisy chain that includes BGA ball, C4 bump, TSV, and μBump.

The package was assembled on a BLR board and tested with and without heat sink. The board is 140 × 383 mm² and 3.42 mm thick. It has 28 metal layers with high-speed, low-loss dielectric material. Board design, assembly, and test are based on IPC-9710. The BLR test was done using TC1 Condition from 0°C to 100°C with 10 minutes of ramp up and cool down along with 10 minutes of dwell time. One complete cycle takes ~40 minutes.

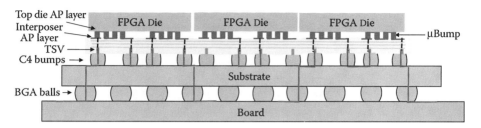

Top die AP layer
Interposer
AP layer
TSV
C4 bumps
BGA balls
µBump
FPGA Die
FPGA Die
FPGA Die
Substrate
Board

FIGURE 3.28
20 nm 3D-IC board-level reliability (BLR) test vehicle.

First failure without heat sink occurred at 8619 cycles and characteristic life (63.2% failure) was reached at 10,792 cycles. Dye and pry FA on the first failed unit indicated the failure was caused by solder ball crack at package corner ball, as shown in Figure 3.29. All failed units have first failure at corner solder balls and not on C4 bump or µBump. This agreed with simulation that the X-interposer package has similar BLR characteristic as a large die flip chip package where corner balls fail early and should be

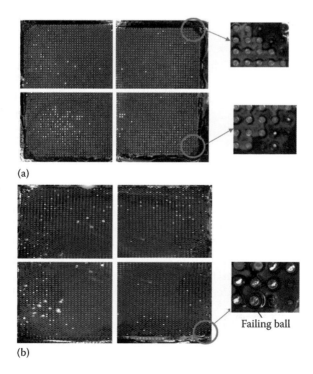

(a)

(b)

Failing ball

FIGURE 3.29
(a) Passing sample, no corner ball crack (with six corner depopulated pads). (b) Failed sample with corner ball crack (with six corner depopulated pads).

considered critical. Including heat sink, onset of first failure was sooner at 5092 cycles and characteristic life was reached at 7565 cycles. However failure mode remained the same, that is, corner BGA balls failed and there was no impact to elements within the package such as µBump, C4, FPGA die ELK, interposer, TSV, and substrate.

The package was also assembled on a Shock and Bend board and tested without heat sink. The board is 185 × 185 mm² and 3.2 mm thick. It has 16 metal layers with high-speed, low-loss dielectric material. Board design, assembly, and test are based on JESD22-B110A for Shock and JESD22-B113 for Bend. The failure criteria for Shock are 10% increase in daisy-chain net resistance and for Bend test is 20% net resistance increase. Results showed Shock test passed both 100G of Condition C and an extended test of 125G; for Bend, the global strain to failure ranged from 1775 to 2359 microstrains.

3.3.3 Three-Dimensional Integration Chip Reliability Look-Ahead Assessments

Reliability is always a significant and inseparable part of technology development. Therefore Xilinx has been constantly running iterations of reliability assessments to discover and to improve any hidden defect, technology weakness, or testing/screening coverage issues. These reliability assessments are especially important during 3D-IC development.

The main 3D-IC reliability assessment (component level), which include temperature cycling (TC, −55°C ~125°C), high-temperature storage (HTS, 150°C), and unbiased highly accelerated test condition-A (uHAST, 130°C/85%RH) with MSL-4 preconditioning (30°C/60%RH/96 hr + 245°C Reflow×3), was applied before these reliability tests.

Figure 3.30a illustrates a 28 nm µBump HTS failure at 1000 hours SEM cross section, significant Sn sidewall wetting, and voiding. The 28 nm µBump

 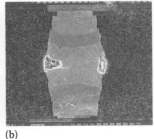

(a)   (b)

FIGURE 3.30
HTS 150°C 1000 hours failing µBump shown in (a) and an improved one shown in (b) (post HTS 150°C 4000 hours of stress).

process was then further improved from metallurgical perspective to enhance the robustness. With the improved μBump, HTS 4000 hours passed without electrical failure (although minor void was observed [Figure 3.30b]).

3.3.3.1 Continuous Process Yield Improvements

KGD is a critical issue for all multidie integration. Xilinx has built-in test capability, thus its 3D-IC top dies are fully tested at wafer level, similar to their monolithic dies done normally. However it is more challenge to do built-in test in passive silicon interposer KGD. A new approach connects the power and ground network in the interposer die into a giant open-short test network in scribe area enabling *almost KGD* test capability. After dicing, these open-short test network return to their normal power and ground function (Figure 3.31).

In addition to design for testability (DFT), redundancy and self-repairing schemes have also been developed [14]. Due to these yield improvement efforts, both 28 and 20 nm largest FPGA (each having ~200,000 and ~600,000 μBumps, respectively) yielded well.

Figure 3.32 illustrated another example of C4 joining yield improvement. Dummy dies were included in chip-on-wafer (CoW) integration, which results in ~10%–15% C4 joining yield gain in a 28 nm heterogeneous 3D-IC product (XC7VH870T). The C4 bridging (caused by warpage of the CoW) at corners of the interposer was then eliminated.

Three generations of Xilinx's FPGA products (28, 20, and 16 nm) are now shipping in production volumes with their defect density approaching monolithic level. Comprehensive reliability tests well beyond JEDEC have been conducted to understand new failure modes and solutions. All of these are due to an effective test vehicle and development strategy.

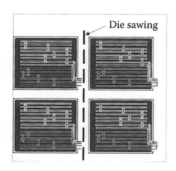

FIGURE 3.31
Plan view of interposer die highlighted with power-ground lines.

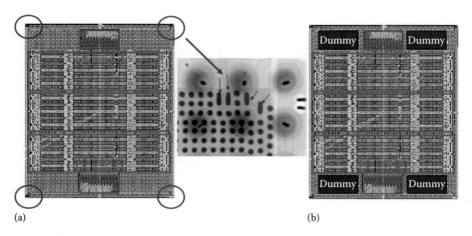

FIGURE 3.32
~10%–15% C4 bridging at four corners of the XC7VH870T was detected (a). The C4 bridging was eliminated after dummy dies included in CoW integration (b).

3.4 Potential Three-Dimensional Integration Chip Future Challenges

3D-IC will be getting more tractions and acceptance from the industry in future. One reason is because 3D-IC technology is reaching maturity, whereas other driving forces and advanced silicon technology continuous scaling trend are becoming more challenging and more expensive.

The challenges expected and which the industry should resolve soon to penetrate high-volume mainstream adoption are given in the following:

1. *The cost-down tendency and margin of the 3D-IC*: It will become a key factor to *promote* 3D-IC adoption by new markets other than emulation, hyperscale data center, and networking.

2. *The 3D-IC capacity and/or supply chain expansion to keep pace with the adoption demand*: Today's limited capacity is expected to be consumed by one or two major fabless design houses. More capacity investment is pursued by foundry and OSATs but the pace needs to accelerate.

3. *The need for an innovative/effective 3D-IC development learning vehicle*: How would it be possible to define a test vehicle as close as to the future 3D-IC product in the development period of a new technology node?

For vertical stacking integration the challenges are more pronounced to identify the technology, design/architecture, and power/thermal space. In the future with advanced silicon technology scaling only becoming more challenging not only from technical but an economic standpoint, it is expected that 3D vertical stacking would gain more interest and momentum in the industry.

Acknowledgment

The authors would like to express our appreciation for the great support from Xilinx's architecture, design, reliability, and FA teams and our partners TSMC's 3D-IC development team, TSMC BTSD team, and SPIL CRD Team. Particular thanks go to our management team for their full commitment and support right throughout the program.

References

1. K. Saban, Xilinx stacked silicon interconnect technology delivers breakthrough FPGA capacity, bandwidth, and power efficiency, Xilinx White Paper: Virtex-7 FPGAs, WP380 (Initial v1.0) October 2011 and (updated v1.2), December 11, 2012.
2. L. Madden, E. Wu, N. Kim, B. Banijamali, K. Abugharbieh, S. Ramalingam, and X. Wu, Advancing high performance heterogeneous integration through die stacking, in *Proceedings of the IEEE 38th European Solid-State Circuits Conference (ESSCIRC)*, Bordeaux, France, 2012, pp. 18–24.
3. C. Erdmann, D. Lowney, A. Lynam, A. Keady, J. McGrath, E. Cullen, D. Breathnach et al., A heterogeneous 3D-IC consisting of two 28 nm FPGA die and 32 reconfigurable high-performance data converters, *IEEE Journal of Solid-State Circuits*, 50(1): 258–269, 2015.
4. S. Wu, J. Wei, and H. Lam, A new SPICE simulation approach for 3D IC integration, SNUG 2013.
5. N. Kim, D. Wu, D.W. Kim, A. Rahman, and P. Wu, Interposer design optimization for high frequency signal transmission in passive and active interposer using through silicon via (TSV), in *Proceedings of the 2011 IEEE 61st Electronic Components and Technology Conference (ECTC)*, Lake Buena Vista, FL, 2011, pp. 1160–1167.
6. N. Kim, D. Wu, J. Carrel, J.H. Kim, and P. Wu, Channel design methodology for 28Gb/s SerDes FPGA applications with stacked silicon interconnect technology, in *Proceedings of the 2012 IEEE 62nd Electronic Components and Technology Conference (ECTC)*, San Diego, CA, 2012, pp. 1786–1793.

7. B. Banijamali, S. Ramalingam, K. Nagarajan, and R. Chaware, Advanced reliability study of TSV interposers and interconnects for the 28nm technology FPGA, in *Proceedings of the 2011 IEEE 61st Electronic Components and Technology Conference (ECTC)*, Lake Buena Vista, FL, 2011, pp. 285–290.

8. Xilinx, Inc., All programmable 3D ICs, http://www.xilinx.com/products/silicon-devices/3dic.html.

9. B. Banijamali, H. Liu, S. Ramalingam, I. Barber, T. Lee, J. Chang, M. Kim, and L. Yip, Reliability evaluation of an extreme TSV interposer and interconnects for the 20 nm technology CoWoS™ IC-package, in *Proceedings of the IEEE Electronic Components and Technology Conference (ECTC)*, San Diego, CA, May 26–29, 2015, pp. 276–280.

10. G. Hariharan, L. Yip, R. Chaware, I. Singh, M. Shen, K. Ng, and A. Xu, Reliability evaluations on 3D IC Package beyond JEDEC, in *Proceedings of the IEEE 67th Electronic Components and Technology Conference (ECTC)*, Lake Buena Vista, FL, 2017, pp. 1517–1522.

11. R. Chaware, G. Hariharan, J. Lin, I. Singh, G. O'Rourke, K. Ng, and S. Pai, Assembly challenges in developing 3D IC package with ultra high yield and high reliability, in *Proceedings of the IEEE Electronic Components and Technology Conference (ECTC)*, San Diego, CA, May 26–29, 2015, pp. 1147–1451.

12. L. Lin, Reliability characterization of chip-on-wafer-on-substrate (CoWoS) 3D IC integration technology, in *Proceedings of the IEEE Electronic Components and Technology Conference (ECTC)*, Las Vegas, NV, May 28–31, 2013, pp. 366–371.

13. W.S. Kwon, M. Kim, J. Chang, S. Ramalingam, L. Madden, G. Tsai, S. Tseng, J.Y. Lai, T. Lu, and S. Chiu, Enabling a manufacturable 3D technologies and ecosystem using 28 nm FPGA with stack silicon interconnect technology, in *Proceedings of the 46th International Symposium on Microelectronics (IMAPS)*, Orlando, FL, 2013, pp. 217–222.

14. R.C. Camarota, J. Wong, H. Liu, and P. McGuire, Applying a redundancy scheme to address post-assembly yield loss in 3D FPGAs, in *Digest of Technical Papers, 2014 Symposium on VLSI Technology*, Hawaii, HI, 2014, pp. 150–151.

4

Challenges in 3D Integration

M. Koyanagi, T. Fukushima, and T. Tanaka

CONTENTS

4.1 Introduction

The 3D integration technology using through-silicon via (TSV) has significantly progressed for these years as represented by 3D-stacked dynamic random-access memory (DRAM) such as hybrid memory cube (HMC) and high bandwidth memory (HBM) [1,2]. The 3D-stacked structure is also employed in a complementary metal–oxide–semiconductor (CMOS) image sensor (CIS) [3]. In addition to these 3D-stacked DRAM and 3D-stacked image sensor, heterogeneous 3D integration technology has increasingly attracted much attention as it is indispensable for future Internet of Things (IoT). In a heterogeneous 3D integration technology, different kinds of chips such as microelectromechanical systems (MEMS), sensor, photonic device chip, and spintronic device are stacked on CMOS chips. Low-power consumption, small form factor, and multifunctionality are required for embedded devices in IoT. Heterogeneous 3D integration can provide these embedded devices with low-power consumption, small form factor, and multifunctionality. We have developed new heterogeneous 3D integration and system integration technologies using self-assembly to achieve multichip-to-wafer stacking with high throughput and high precision. In this chapter, these new heterogeneous 3D integration and system integration technologies for IoT are described.

4.2 Past Challenges in Three-Dimensional Integration

The first 3D large-scale integration (LSI) test chip having three device layers was reported by Akasaka and Nishimura [4]. In this 3D LSI test chip, metal–oxide–semiconductor field-effect transistors (MOSFETs) are fabricated in poly-Si films deposited on an LSI wafer. The poly-Si film is recrystallized by laser annealing before fabricating the MOSFETs to improve the electrical characteristics of the poly-Si film. A 3D LSI fabrication process using a thinned-wafer transfer method was proposed by Hayashi et al. [5]. In this 3D LSI process, an LSI wafer is glued to a supporting material and then the LSI wafer is thinned from the back surface by mechanical grinding and polishing to a thickness of approximately 0.5 μm. This thinned LSI wafer is directly bonded to another LSI wafer and then the supporting material is removed from the thinned LSI wafer. To avoid the transfer of extremely thinned wafer, we proposed a 3D LSI with vertical interconnection called a TSV as shown in Figure 4.1a in 1989 [6–9]. We have developed four kinds of TSVs so far as shown in Figure 4.1b. We fabricated various 3D prototype chips using poly-Si TSV, W-TSV, and Cu-TSV. Our first 3D prototype chip is a 3D-stacked image sensor with poly-Si TSV fabricated in 1999 as shown in Figure 4.2 [10]. Furthermore we fabricated a 3D-stacked memory in 2000, 3D-stacked artificial retina chip in 2001, and 3D-stacked processor in 2002 using poly-Si TSV [11–13].

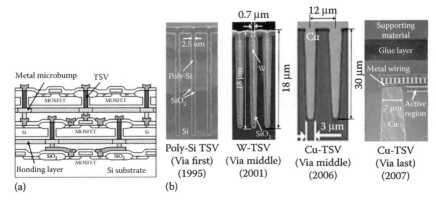

FIGURE 4.1
Cross-sectional structure of 3D LSI with TSVs (a) and cross-sectional view of TSVs (b).

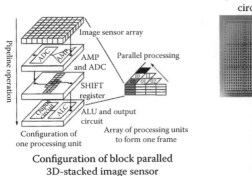

First layer (Photosensor circuit)

Second layer (Register circuit)

Third layer (ADC and ALU circuit)

Image sensor array

Pipeline operation

ADC AMP

AMP and ADC

Parallel processing

SHIFT register

Chopper circuit ALC

ALU and output circuit

Configuration of one processing unit

Array of processing units to form one frame

Configuration of block paralled 3D-stacked image sensor

Quartz class

Stacked chips

Si interposer

Photo of 3D-stacked image sensor

FIGURE 4.2
Photomicrographs of fabricated 3D-stacked image sensor test chip with poly-Si TSVs.

4.3 Challenges in Three-Dimensional System Integration

We can create new intelligent system modules for not only high-end information system application but also mobile and consumer applications by employing the 3D integration technology [14,15]. A typical example of these intelligent system modules is a 3D-stacked image sensor system module for advanced driver assistance systems (ADAS) as shown in Figure 4.3 where two 3D-stacked image sensors and 3D-stacked multicore processors are integrated on a silicon interposer [16,17].

A prototype 3D-stacked CMOS image sensor with the resolution of quarter video graphics array (QVGA) (320 × 240 pixels) and the frame rate of 10,000 frames/s was fabricated by the 3D integration technology with the backside-via [18]. A configuration of a prototype 3D-stacked CMOS image sensor with a block-parallel architecture that contains a number of image signal-processing elements is shown in Figure 4.4 [19–21]. The 3D-stacked CMOS image sensor consists of four layers of image sensor, correlated double sampling (CDS) with programmable gain amplifier (PGA) array, analog-to-digital converter (ADC) array, and interface circuit array. The CMOS image sensor (CIS) layer and CDS layer were designed with a standard 0.18 µm CMOS image sensor and mixed signal technologies, respectively. We have employed a block-parallel architecture for ADC to convert a large amount of analog data from image sensors to digital data with high speed. This block-parallel ADC architecture employs a 9-bit time-interleaved successive approximation (SAR)

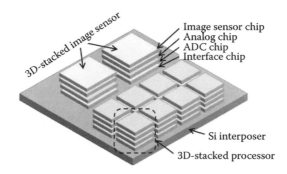

FIGURE 4.3
Configuration of 3D-stacked image sensor system module for advanced driver assistance systems (ADAS).

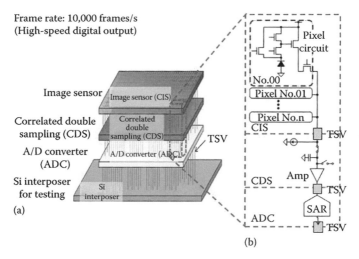

FIGURE 4.4
Configuration of one image signal processing element in 3D-stacked image sensor. (a) Structure of stacked image sensor and (b) configuration of one circuit block.

ADC with digital noise cancellation circuit. The ADC was designed with a standard 90-nm 1-Poly 9-Metal CMOS technology [20]. The three-dimensional structure of designed 3D-stacked image sensor is depicted in Figure 4.5a. Figure 4.5b shows X-ray CT scan image and SEM cross-sectional view of fabricated 3D-stacked image sensor. It is clearly seen in the X-ray CT scan image that four layers with many TSVs are vertically stacked. The die size of each layer is 5×5 mm^2 and each layer has approximately two thousands of TSVs. The thickness of Si substrate and the diameter of TSV are approximately 50 and 5 μm, respectively. We confirmed the successful operation of fabricated prototype 3D-stacked image sensor.

A prototype 3D-stacked dependable multicore processor was fabricated by the 3D integration technology with the backside-via [22–23]. The conceptual

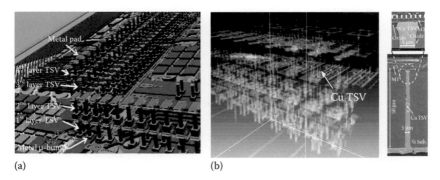

(a)   (b)

FIGURE 4.5
Three-dimensional display of designed 3D-stacked image sensor by computer graphics (a) and X-ray CT scan image and SEM cross-sectional view of fabricated 3D-stacked image sensor (b).

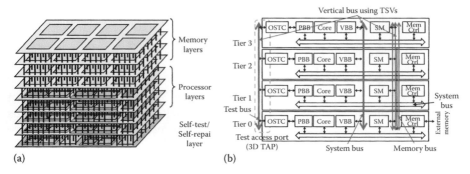

(a)   (b)

FIGURE 4.6
Conceptual structure (a) and functional circuit block diagram (b) of 3D-stacked multicore processor with self-test and self-repair function.

structure of this 3D-stacked multicore processor is illustrated in Figure 4.6. A circuit block diagram and a die photo of core processor fabricated by a standard 90-nm 1-Poly 9-Metal CMOS technology are shown in Figure 4.7a. The die size is $5 \times 5mm^2$. Approximately two thousands of TSVs per each core processor layer were assigned to form the system bus, memory bus, and test bus including TSVs in I/O circuits. This core processor chip exhibited the performance of 350 Mips (Dhrystone 2.1) at a clock frequency of 200 MHz. We fabricated a four-layer stacked multicore processor and a four-layer stacked cache memory. Figure 4.7b shows the X-ray CT scan image and SEM cross-sectional view of fabricated 3D-stacked multicore processor. It is clearly seen in the figure that the stacked structure with many TSVs is successfully formed. We confirmed using internal built-in-self-test (BIST) circuits that the fabricated prototype 3D-stacked multicore processor exhibited excellent characteristics [24].

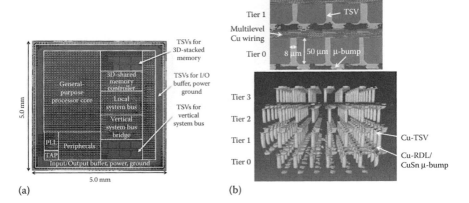

(a)   (b)

FIGURE 4.7
A circuit block diagram and a die photo of core processor (a) and X-ray CT scan image and SEM cross-sectional view of fabricated 3D-stacked multicore processor (b).

4.4 Challenges in Three-Dimensional Heterogeneous Integration

We challenge to achieve a new heterogeneous 3D LSI for IoT application called a superchip as shown in Figure 4.8 by stacking various kinds of chips with different sizes, different devices, and different materials. We have developed a new 3D integration technology as shown in Figure 4.9 to fabricate such a new heterogeneous 3D LSI. We call this new 3D integration technology a reconfigured wafer-to-wafer (R-W2W) 3D integration technology [25–27]. A reconfigured wafer is fabricated by bonding many known good dies (KGDs) onto a carrier wafer and then these reconfigured wafers are bonded in this 3D integration technology. We have developed new self-assembly and electrostatic temporary bonding methods to bond many

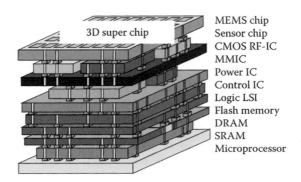

FIGURE 4.8
An example of heterogeneous 3D LSI.

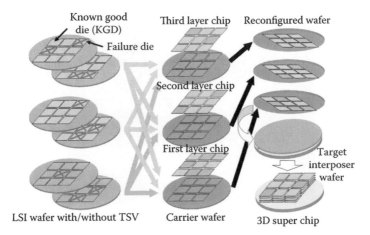

FIGURE 4.9
Production procedure for heterogeneous 3D LSIs in reconfigured wafer-to-wafer (R-W2W) 3D integration technology.

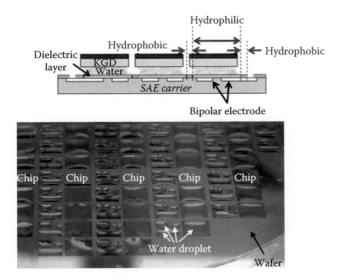

FIGURE 4.10
Concept of self-assembly and electrostatic temporary bonding and a photo of 8-inch wafer with hydrophilic areas and hydrophobic areas after simultaneously supplying liquid droplets with four sizes on hydrophilic areas.

known good dies (KGDs) onto a carrier wafer with high alignment accuracy and high throughput. The surface tension of liquid droplet is utilized in the self-assembly to simultaneously align many dies as shown in Figure 4.10. Hydrophilic areas and hydrophobic areas are formed on the surface of wafer or chip. We have succeeded in simultaneously aligning five hundreds of chips with the average alignment accuracy of 0.5 μm within 0.1 second.

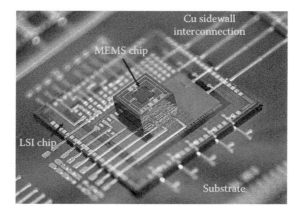

FIGURE 4.11
A photo of heterogeneous 3D LSI test chip where a pressure sensor MEMS chip is integrated on CMOS chip, which is stacked on an active interposer wafer.

So far we have fabricated several heterogeneous 3D LSI test chips by stacking MEMS chip, spin memory chip, and photonic device chip on CMOS chips using self-assembly [28–30]. Figure 4.11 shows a photo of heterogeneous 3D LSI test chip where a pressure sensor MEMS chip is integrated on CMOS chip [28]. The process flow to fabricate 3D LSIs by the R-W2W integration technology is described in Figure 4.12 [31]. Many chips (KGDs) with metal microbumps and nonconductive films (NCFs) are simultaneously face-down bonded onto the carrier wafer after self-assembly. Then these chips are simultaneously thinned from the backside, and metal microbumps are formed on

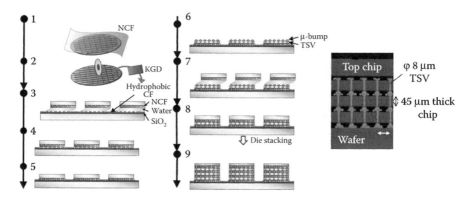

FIGURE 4.12
Fabrication process flow to fabricate 3D LSIs by a reconfigured wafer-to-wafer (R-W2W) integration technology. 1—Lamination, 2—dicing, 3—first-layer die release, 4—self-assembly (alignment), 5—die-bonding, 6—Wafer thinning/TSV and µ-bump formation, 7—second-layer die release, 8—self-assembly and die-bonding, and 9—completed 3D stacking.

the backside surface of chips after thinning. A number of 3D-stacked LSIs can be simultaneously fabricated by repeating this sequence. Cross-sectional photomicrograph of 3D-stacked test chip fabricated according to such a 3D integration technology is also shown in Figure 4.12.

4.5 Challenges toward Future Three-Dimensional Integration

We are challenging to achieve a future high-performance and low-power 3D LSI with high-density TSV and microjoint of more than one million. The pitches of TSV and microjoint will be reduced down to less than 5 μm to achieve such a 3D LSI. The size of microjoint and the diameter of TSV should be less than 2 and 1 μm, respectively. We have proposed to employ W-TSV for a via-middle and Ni-TSV for a via-last to achieve such a fine TSV. We have succeeded to completely fill W into the deep silicon hole with the diameter of 0.5 μm and the aspect ratio of more than 30 by atomic layer deposition (ALD) [32]. We also filled Ni into the deep silicon hole by electroless plating [33]. A metal barrier layer and a seed layer for electroplating are not necessary in W-TSV and Ni-TSV. However, we still need Cu-TSV with larger diameter for power delivery. Therefore, W/Cu or Ni/Cu hybrid TSVs will be employed in a future 3D LSI [34]. To further increase the TSV density in a future 3D LSI, the diameter of TSV for signal propagation should be less than 0.1 μm. A completely new method will be needed to fabricate such a TSV with extremely small diameter. Then we propose a novel method to fabricate a fine TSV with the diameter of less than 0.1 μm using a directed self-assembly (DSA) of diblock copolymer. Figure 4.13 shows the new concept of TSV formation based on advanced DSA with nanocomposites consisting of diblock copolymers and nanosized metal particles [35]. TSVs are formed by the phase separation of block copolymers with simple heating at below 300°C. Nanosized metal particles are included into copolymer with cylindrical structure after the phase separation of copolymers. To confirm the capability to fabricate such a fine TSV using DSA, we filled diblock copolymer with metal nanoparticles into a deep Si trench with the diameter of 3 μm and the depth of 10 μm and confirmed that nano-ordered lamella and cylindrical structures were formed in the diblock copolymer, PS-b-PMMA with the molecular ratio of 1:1 and 2:1, respectively, after the phase separation at 280°C. The PS-b-PMMA with higher molecular weight showed a larger width of the periodic structure of more than 50 nm.

The microjoint with the small size of less than 2 μm is indispensable to fabricate a fine TSV with the diameter of less than 1 μm. A hybrid bonding may be useful to bond two wafers with many microjoint with the small size of less than 2 μm [36]. However it is difficult to employ a conventional hybrid bonding in chip-to-chip bonding and chip-to-wafer bonding. Then we have

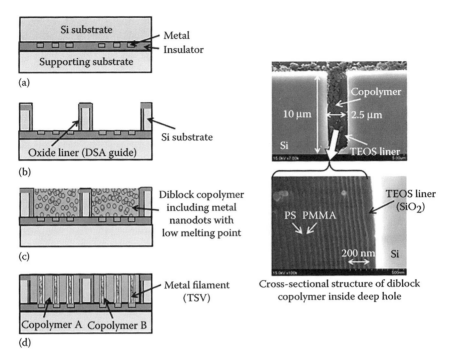

FIGURE 4.13
New concept of TSV formation based on advanced directed self-assembly (DSA) with nano-composites consisting of diblock copolymers and nanosized metal particles. (a) Thinning of Si substrate, (b) formation of Si deep hole (RIE), (c) deposition of diblock copolymer including metal nanodots with low melting point, and (d) nanophase separation.

proposed a new hybrid bonding method for chip-to-chip bonding and chip-to-wafer bonding in which a novel inorganic anisotropic conductive film (i-ACF) comprising ultrahigh density of Cu nanopillar (CNP) and alumina matrix is used. SEM images of i-ACF film with the top view and the cross-sectional view are shown in Figure 4.14. The i-ACF film is formed by anodic oxidation of aluminum film and Cu electroplating [36]. The diameter and pitch of Cu nano-pillars are 60 and 100 nm, respectively. In order to evaluate the electrical characteristics of Cu–Cu joining bonded through the i-ACF film, we fabricated a test chip with a huge number of Cu electrodes of 4.3 million. The size and pitch of Cu electrodes are 3 and 6 μm, respectively. The cross-sectional image of fabricated test chips bonded using i-ACF film is also shown in Figure 4.14. Test chip and interposer chip are electrically connected by ultrahigh-density CNPs. We confirmed that all of 4.3 million Cu joints with the electrode size of 3 μm were completely connected through ultrahigh-density CNPs [37–38]. The joining resistance was approximately 30 mΩ for each pair of joining.

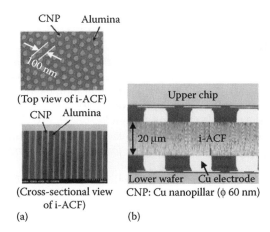

FIGURE 4.14
SEM images of i-ACF film with the top view and the cross-sectional view (a) and cross-sectional image of fabricated test chip bonded using i-ACF film (b).

4.6 Summary

We have developed various 3D integration technologies using TSV and metal microbump for a long time. In addition, we have also developed various 3D system-on-chips and heterogeneous 3D LSIs using self-assembly and electrostatic bonding. Furthermore, we are challenging to develop new technologies such as DSA TSV and CNP hybrid bonding for future 3D LSIs with high-density TSVs and microjoints.

References

1. J. Jeddeloh and B. Keeth, Hybrid memory cube new DRAM architecture increases density and performance, *Digest of Technical Papers, Symposium on VLSI Technology*, pp. 87–88 (2012).
2. D. U. Lee et al., A 1.2V 8Gb 8-channel 128GB/s high-bandwidth memory (HBM) stacked DRAM with effective microbump I/O test methods using 29nm process and TSV, *IEEE International Solid-State Circuits Conference* (ISSCC), pp. 432–433 (2014).
3. S. Sukegawa et al., A 1/4-inch 8Mpixel back-illuminated stacked CMOS image sensor, *IEEE International Solid-State Circuits Conference* (ISSCC), pp. 484–485 (2013).
4. Y. Akasaka and T. Nishimura, Concept and basic technologies for 3-D IC structure, *IEEE International Electron Devices Meeting* (IEDM), pp. 488–491 (1986).

5. Y. Hayashi et al., A new three dimensional IC fabrication technology, stacking thin film dual-CMOS layers, *IEEE International Electron Devices Meeting (IEDM)*, pp. 657–660 (1991).

6. M. Koyanagi, Roadblocks in achieving three-dimensional LSI, *Proceedings of the 8th Symposium on Future Electron Devices*, pp. 50–60 (1989).

7. T. Matsumoto et al., Three-dimensional integration technology based on wafer bonding technique using micro-bumps, *International Conference on Solid State Devices and Materials (SSDM)*, pp. 1073–1074 (1995).

8. T. Matsumoto et al., New three-dimensional wafer bonding technology using the adhesive injection method, *Japanese Journal of Applied Physics*, 1(3B): 1217–1221 (1998).

9. M. Koyanagi et al., Future system-on-silicon LSI chips, *IEEE Micro*, 18(4): 17–22 (1998).

10. H. Kurino et al., Intelligent image sensor chip with three dimensional structure, *IEEE International Electron Devices Meeting (IEDM)*, pp. 879–882 (1999).

11. K. W. Lee et al., Three-dimensional shared memory fabricated using wafer stacking technology, *IEEE International Electron Devices Meeting (IEDM)*, pp. 165–168 (2000).

12. M. Koyanagi et al., Neuromorphic vision chip fabricated using three-dimensional integration technology, *IEEE International Solid-State Circuits Conference (ISSCC)*, pp. 270–271 (2001).

13. T. Ono et al., Three-dimensional processor system fabricated by wafer stacking technology, *IEEE COOL Chips*, pp. 186–193 (2002).

14. M. Koyanagi et al., Three-dimensional integration technology based on wafer bonding with vertical buried interconnections, *IEEE Transactions on Electron Devices*, 53(11): 2799–2808 (2006).

15. M. Koyanagi et al., High-density through silicon vias for 3-D LSIs, *Proceedings of the IEEE*, 97(1): 49–59 (2009).

16. M. Koyanagi, Heterogeneous 3D integration -Technology enabler toward future super-chip, *IEEE International Electron Devices Meeting (IEDM)*, pp. 8–15 (2013).

17. M. Koyanagi, 3D system module with stacked image sensors, stacked memories and stacked processors on a Si interposer, *IEEE International Solid State Circuits Conference (ISSCC) 3D-Forum* (2014).

18. K.-W. Lee et al., Characterization of chip-level hetero-integration technology for high-speed, highly parallel 3D-stacked image processing system, *IEEE International Electron Devices Meeting (IEDM)*, pp. 785–788 (2012).

19. K. Kiyoyama et al., A block-parallel SAR ADC for CMOS image sensor with 3-D stacked structure, *International Conference on Solid State Devices and Materials (SSDM)*, pp. 1055–1056 (2011).

20. K. Kiyoyama et al., Very low area ADC for 3-D stacked CMOS image processing system, *IEEE International 3D System Integration Conference (3DIC)*, p. 5.1 (2012).

21. K. Kiyoyama et al., A block-parallel ADC with digital noise cancelling for 3-D stacked CMOS image sensor, *IEEE International 3D System Integration Conference (3DIC)* (2013).

22. T. Fukushima et al., New heterogeneous multi-chip module integration technology using self-assembly method, *IEEE International Electron Devices Meeting (IEDM)*, pp. 499–502 (2008).

23. T. Fukushima et al., New chip-to-wafer 3D integration technology using hybrid self-assembly and electrostatic temporary bonding, *IEEE International Electron Devices Meeting (IEDM)*, pp. 789–792 (2012).

24. H. Hashimoto et al., Highly efficient TSV repair technology for resilient 3-D stacked multicore processor system, *IEEE International 3D System Integration Conference (3DIC)* (2013).

25. T. Fukushima et al., New three-dimensional integration technology based on reconfigured wafer-on-wafer bonding technique, *IEEE International Electron Devices Meeting (IEDM)*, pp. 985–988 (2007).

26. T. Fukushima et al., Three-dimensional integration technology based on reconfigured wafer-to-wafer and multichip-to-wafer stacking using self-assembly method, *IEEE International Electron Devices Meeting (IEDM)*, pp. 349–352 (2009).

27. T. Fukushima et al., Self-assembly technology for reconfigured wafer-to-wafer 3D integration, *IEEE Electronics Components and Technology Conference (ECTC)*, pp. 1050–1053 (2010).

28. K.-W. Lee et al., A cavity chip interconnection technology for thick MEMS chip integration in MEMS-LSI multichip module, *Journal of Microelectromechanical Systems*, 19(6): 1284–1291 (2010).

29. K.-W. Lee et al., Three-dimensional hybrid integration technology of CMOS, MEMS, and photonics circuits for optoelectronic heterogeneous integrated systems, *IEEE Transactions on Electron Devices*, 58(3): 748–757 (2011).

30. T. Tanaka et al., Ultrafast parallel reconfiguration of 3D-stacked reconfigurable spin logic chip with On-chip SPRAM (spin-transfer torque RAM), *Symposia on VLSI Technology and Circuits (VLSI2012)*, pp. 169–170 (2012).

31. Y. Ito et al., Development of highly-reliable microbump bonding technology using self-assembly of NCF-covered KGDs and multi-layer 3D stacking challenges, *IEEE Electronics Components and Technology Conference (ECTC)*, pp. 336–341 (2015).

32. H. Kikuchi et al., Tungsten through-silicon via technology for three-dimensional LSIs, *Japanese Journal of Applied Physics*, 47(4): 2801–2806 (2008).

33. K.-W. Lee et al., Effects of electro-less Ni layer as barrier/seed layers for high reliable and low cost Cu TSV, *International 3D System Integration Conference (3DIC)* (2014).

34. M. Murugesan et al., High density 3D LSI technology using W/Cu hybrid TSVs, *IEEE International Electron Devices Meeting (IEDM)*, pp. 139–142 (2011).

35. T. Fukushima et al., New concept of TSV formation methodology using directed self-assembly (DSA), *IEEE International 3D System Integration Conference (3DIC)* (2016).

36. K. Yamashita et al., Copper-filled anodized aluminum oxide -a potential material for low temperature bonding for 3D packaging, *ICEP- IAAC*, pp. 571–574 (2015).

37. K.-W. Lee et al., Novel reconfigured wafer-to-wafer (W2W) hybrid bonding technology using ultra-high density nano-Cu filaments for exascale 2.5D/3D integration, *IEEE International Electron Devices Meeting (IEDM)*, pp. 185–188 (2015).

38. K.-W. Lee et al., Novel W2W/C2W hybrid bonding technology with high stacking yield using ultra- fine size, ultra-high density Cu nano-pillar (CNP) for exascale 2.5D/3D integration, *IEEE Electronics Components and Technical Conference (ECTC)*, pp. 350–355 (2016).

5

Wafer-Level Three-Dimensional Integration Using Bumpless Interconnects and Ultrathinning

Takayuki Ohba

CONTENTS

5.1 Introduction

Integrated circuits (ICs) based on planar technology started in the 1960s and have led to today's huge semiconductor industry and will be a key technology in realizing a global infrastructure for the Internet-of-Things (IoT) and the Internet-of-Everything (IoE) [1]. Twenty years after the advent of ICs, the concept of three-dimensional integration (3DI) was proposed. The beginning of 3DI was the so-called transistor-based 3DI in the front-end-of-line (FEOL) for stacking complementary metal–oxide–semiconductor (CMOS) devices (a monolithic-stacked structure of n-MOSFETs [metal–oxide–semiconductor field-effect transistor] and p-MOSFETs) to fabricate high-density ICs. In 2000s, packaging-based 3DI, such as chip-on-board (COB) and chip-on-chip (COC) with wire bonding, was developed to fabricate high-performance electronics. A typical product is a system-in-package (SiP) consisting of a stack of 4–5 chips for mobile applications. Due to these two different approaches, that is, transistor- or packaging-based 3DI, some technical misunderstandings often occurred when people heard the term 3DI technology. For instance, current interconnecting ICs in back-end-of-line (BEOL) for microprocessing units (MPUs) or graphics processing units (GPUs) were realized by 12–14 level Cu/low-k multilevel interconnects. This is obviously a 3D structure and comes from the requirements for high device performance and density. Packaging-based 3DI consists of a chip-based stack after singulation of the wafer, and many fabrication methods have been introduced leading to complication and confusion in attempts to classify 3DI technologies. In any case, attempts to apply the early generation of 3DI were delayed. This is because conventional miniaturization, involving scaling in planar and/or 2D processes, had succeeded in reaching the submicron level, enabled by inexpensive ICs with high density and high capacity. Wafer enlargement, such as from 8 to 12 inch in diameter, also helped to reduce chip costs.

Interest in packaging-based 3DI technology using wafer-level processing has been increasing again. This is driven by the physical and economic limits of conventional scaling, which is no longer a main stream for the increasing demands for device performance, system form factor, and total manufacturing cost. This chapter discusses a BEOL-compatible wafer-level 3DI approach, including state-of-the-art interconnecting technologies such as bumpless vertical interconnects and ultrathinning of 300 mm wafers down to micrometer thickness [2–4].

5.2 Co-Engineering by 3D and 2D

5.2.1 Delay of Three-Dimensional Integration Technology

The two-dimensional (2D) semiconductor industry and its growth are supported by mature scaling technologies following Moore's empirical law [5]: a doubling transistor density every 18 months by scaling. To discuss 3D and to compare chip-based packaging, utilizing wafers for IC manufacturing was a significant difference and it made it easy to increase production volumes. Thus, scaling and wafer processing are the key factors for improving performance, shrinking chip size, and reducing costs, simultaneously. Although chip-based manufacturing has an advantage in the case of small production volumes and short cycle-time from design to production, most chip-based processes, including tools and materials, were not compatible to that of wafer based and need to adjust to the specifications for volume production. Those facility trends were another reason for the delay in adaption of 3DI processes.

5.2.2 Economic and Technical Issues for Lithography

Two-dimensional scaling will soon face a severe economic crisis due to the expensive lithography processes and facilities required. Reducing costs requires the adoption of advanced lithography technologies, which, together with peripheral support facilities, account for one-third to one-fourth of the total cost of a manufacturing line. In short, although useful for reducing chip size, scaling is extremely burdensome in terms of capital investment. Large-scale investments have so far been made considering the technologies that will be available two to three generations ahead, for example, 10 nm technology should be applied to nodes of <5 nm, which will also face physical limitations. This is a business scenario based on the empirical rule that profits are made several generations after investments, for reasons involving the tradeoffs between product sales and facility depreciation.

According to this empirical rule, an investment in 15 nm technology needs to be made in consideration of its applicability to 10, 5, and 3 nm technologies. However, the price of extreme ultraviolet (EUV; $\lambda = 13.5$ nm) lithography machines is >100 million USD, which is more than twice that of existing ArF immersion lithography machines, and their current throughput is around one-tenth or less. When converted into the processing capacity of a current large-scale fabrication facility (e.g., 50,000 incoming wafers per month), based on this system performance, an investment of approximately 2 billion USD will be required for EUV technology. Assuming that the past lifelong sales for each generation are approximately 10 times the corresponding business investment, the corresponding market size necessary for this investment is more than 20 billion USD. Based on the 300 billion USD, total worldwide semiconductor market, this expected market size for one product and one manufacturer is not realistic. In short, this is the limit of 2D scaling in light of the economics of the industry, and it is difficult to find a scenario of victory at present.

5.2.3 Co-Engineering Using Three Dimensional for Next Generation of Manufacturing

3D stacking, in combination with improved I/O signals of 2D, is anticipated to overcome these scaling problems [6–8]. In the case of globally interconnecting large chips with high density, the signal traffic latency increases with an increase in the physical length (L_{2D}) from block-to-block as well as logic-to-memory due to wire resistance–parasitic capacity (RC) delay, as shown in Figure 5.1. This latency will be reduced by 3D-stacked structure when the physical length (L_{3D}) through one chip to the chip below becomes shorter than L_{2D}. The shorter interconnects ($L_{3D} \ll L_{2D}$) are also advantageous for power reduction, noise reduction, and clock cycle propagation across the entire chip. Therefore, how to shrink the vertical length of 3D interconnects and how to increase their density will be key factors in developing future technology roadmaps, instead of the features of 2D scaling technology.

Vertical interconnects without bump electrodes between chips, such as Cu BEOL processes, provide the lowest RC characteristics because of the

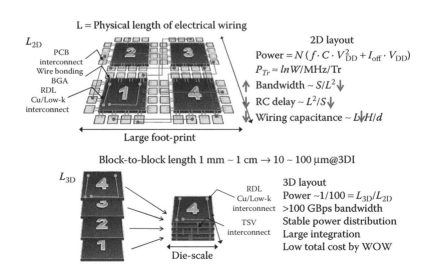

FIGURE 5.1
A comparison of wiring length and layout for 2D and 3D chip sets. Miniaturizing the layout using 3DI provides low-power consumption, higher bandwidth, and higher integration.

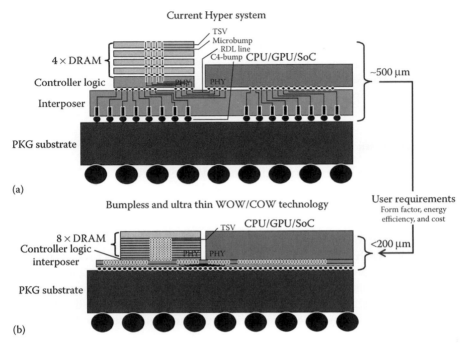

FIGURE 5.2

A comparison of (a) bump and (b) bumpless interconnects using TSVs for 3D memory stack structures, assuming memory stacks containing four memories + controller (five stacks) and eight memories + controller (nine stacks). Bumpless interconnects used for the ultrathin WOW process can be formed with higher density (narrower pitch) compared with TSVs and bumps due to the limitations of bump size and pitch.

RC reduction by no bump structure. Figure 5.2 shows a comparison of bump and bumpless interconnects using through-silicon vias (TSVs) for 3D memory–memory-stacked structures. 3D memory is implemented by placing memories and MPUs side-by-side on an interposer. In the case of bumpless and ultrathin processes, the vertical length of interconnects from die to die is 1/10th compared with bumps, assuming a bump height of 30 μm and a die thickness of 70 μm. Even for stack containing two times more memory chips, the total height of the 3D memory will be a half of the height of the conventional approach.

Wafer-level 3DI for such high-capacity memories has productivity benefits compared with other 3D methods such as COC and chip-on-wafer (COW). Chip-based processes need a pick-and-place process (~2 s/chip) after wafer singulation and thus suffer from low throughput. When the cycle-time in pick-and-place is reduced, misalignment of chip placement typically increases by more than 5 μm, which limits the density of TSVs.

A stack containing three 300 mm wafers provides a total silicon surface larger than that of a single 450 mm wafer. This is important in considering whether wafer enlargement should be continued, taking into account the cost benefits of increasing wafer sizes to 450 mm compared with utilizing 300 mm wafers. Moreover, retaining the standard 300 mm wafer size for stacking ensures compatibility with existing manufacturing facilities in front-end processing and helps utilize the mature process technology that has been developed for wafer processing.

5.3 Bumpless Interconnecting and Wafer-Level Three-Dimensional Integration

5.3.1 Overview of Bumpless Interconnects

Bumpless interconnects using TSVs are a second-generation alternative to the use of TSVs with microbumps. The bumpless interconnects process involves a *thinning-first* process before bonding wafers, followed by a *via-last* process, meaning that interconnects are formed after bonding the wafers, as shown in Figure 5.3. This method is compatible with the BEOL, similar to multi-level metallization in which dielectric deposition using thinned wafers and

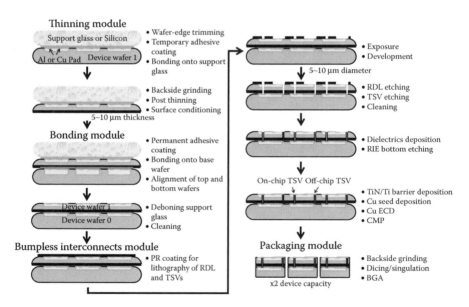

FIGURE 5.3

Bumpless interconnects using ultrathinning, TSVs, and wafer-on-wafer (WOW) process flow. Vertical interconnects for TSVs are formed after wafer bonding from the front side. Additional wafers can be stacked on top without any limitation on the number of wafers. These modules can also be applied to chip-on-wafer (COW) integration. On-chip and off-chip TSV, respectively, represent bumpless interconnects formed in the device area and the area around devices, including gap fill materials in COW.

Al and/or Cu metallization is replaced with bumpless interconnects using TSVs. After wafer thinning and bonding, via-hole etching is carried out on a silicon substrate from the front surface of the wafer having the device layer and a dielectric layer for the multilevel interconnects. In principle, any number of thinned 300 mm wafers can be stacked to fabricate a large-capacity memory device. In the case of a heterogeneous stack using different chip sizes such as memory, logic, and passive device, COW can be used. For instance, after preparing a redistribution line (RDL) on an MPU wafer, the RDL pad layout is adjusted to match the dynamic random-access memory (DRAM) I/Os, and the DRAM chip is picked up and placed on the MPU wafer, so that the chip and wafer are interconnected by bumpless TSVs. For a 3D stack memory, a side-by-side array with an MPU is possible, where a 3D memory and MPU are stacked in a face-to-face manner onto an interposing 300 mm wafer. By applying a wafer-level packaging process for the front side, the interposing wafer is thinned down to <5 µm, and then TSV and RDL processes are carried out on the thinned silicon wafer, as shown in Figure 5.2b.

The development of WOW has proceeded through four modules, classified along the process flow. The modules include (1) a thinning module for thinning the wafer substrates in which devices are implemented, (2) a stacking module for bonding and stacking the wafers, (3) a TSV interconnects module for forming Cu interconnects embedded in upper and lower wafers with TSVs, and (4) a packaging module for singulating the stacked wafers. Dual-damascene interconnects form (RDL) and also serve as a counter electrode for the subsequent stacked wafer.

5.3.2 Thickness of Wafer and Ultrathinning

The thickness of the wafer is a critical dimension for the aspect ratio of TSVs because the aspect ratio is determined by the diameter and the wafer thickness. As a thinned wafer has low rigidity caused by bowing and/or warpage due to a mismatch in the coefficients of thermal expansion (CTE) between silicon and dielectrics, there is a thickness limit of ~50 µm at which standalone thinned wafers can be safely handled. In our process, a wafer was bonded to a support substrate (glass or silicon) using a temporary bonding material, and thinning was carried out from the back side. The thinned wafer was bonded to another wafer by back-to-face bonding using a permanent bonding material, and then the support substrate was removed. Thus, there is no need for high rigidity even with a thin wafer.

In our recent work, the typical thickness of thinned wafers ranged from 5 to 10 µm. When the thicknesses of the device layers in a DRAM and an MPU are assumed to be approximately 5 and 10 µm, respectively, the aspect ratio (depth-to-diameter ratio) of a TSV is only 2–4 for a TSV diameter ranging from 5 to 10 µm, whereas conventional TSVs for a 50 µm wafer thickness have aspect ratios ranging from 5 to 10. With decreasing aspect ratio, the processing time in the TSV processes, such as etching, thin film deposition, and

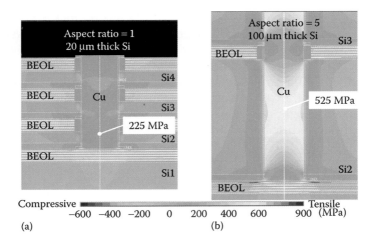

FIGURE 5.4
Stress simulation using the finite element method (FEM) for Si thicknesses of 100 (b) and 20 μm (a) after stacking three wafers and forming a TSV with a diameter of 30 μm. A 10 μm Cu/low-k BEOL layer is formed on every wafer surface, and thus the depths of the TSVs are 110 and 30 μm, respectively.

metal filling, decreases to about 1/5 at most, and the step coverage significantly improves. A reduced TSV length is also considerably advantageous for data transfer and power distribution with high-energy efficiency.

In addition, with the use of small TSVs, stress induced by a mismatch in the CTE between Cu and Si decreases with decreasing aspect ratio of the TSV, as shown in Figure 5.4 [9]. Stress at the center of a Cu plug in a 100 μm thick wafer is 525 MPa and decreases to 225 MPa in a 20 μm wafer. The small aspect ratio provided by an ultrathin wafer also has the advantage of reducing stresses generated in the silicon itself, in the bottom and top Cu-TSV, and in interface regions having different CTEs.

5.4 Details of Wafer-Level Three-Dimensional Process

5.4.1 Thinning Module

A wafer is bonded to a support substrate in advance with a temporary adhesive (thermoplastic resin). Thinning is performed by grinding (Back Grind, or BG) within several micrometers of the target thickness, followed by polishing until the final thickness is achieved. The thinned wafer is permanently bonded (thermosetting resin) to the device surface of another wafer, and then the support substrate is removed. In the case of *face-to-face* WOW employing bumps between wafers, the wafer thickness is limited by its poor rigidity and the critical thickness is 50–100 μm. By using bumpless and *back-to-face* WOW processes, the wafer thickness is limited mainly by the mechanical precision of the grinding. With optimization of mechanical parallelism during grinding, a thickness precision of 2 µm in a 300 mm-diameter DRAM wafer has been achieved [10]. A 300 mm wafer on which 32 nm-node SDRAM devices (2GB DDR3) are fabricated can be thinned down to 4 µm with a total thickness variation (TTV) of around 1 µm, which is just 0.5% of the initial thickness [11]. With a wafer thickness of 4 µm, visible light starts to pass through the wafer. Remarkably, no degradation of retention time before and after thinning was observed, even in a Si wafer with a thickness of 4 µm.

5.4.2 Stacking Module

An organic adhesive such as benzo-cyclo-butene (BCB) polymer [12–14] and/ or an alternative thermosetting resin with a thickness of approximately <5 µm was used to permanently bond the wafer. BCB adhesives start to polymerize with increasing temperature and are solidified at temperatures of 200°C–250°C. For WOW, the wafers are aligned just before being permanently bonded. To ensure appropriate alignment, infrared light passing through the silicon substrate is used. Wafers to be bonded to one another in WOW are originally thin and are therefore highly transmissive. With a low-temperature bonding process and an optimized curing duration and viscosity, the average misalignment between wafers can be made as small as several micrometers as shown in Figure 5.5. This is predominantly due to the bonding temperature, that is, reducing the bonding temperature provides smaller misalignment. On the other hand, because any gaseous solvent

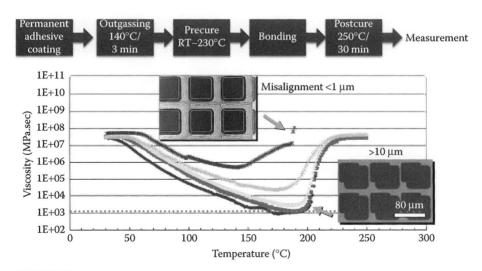

FIGURE 5.5
Relationship of misalignment of wafer bonding and viscosity of permanent adhesive material as a function of temperature.

escaping from the adhesive after the bonding process would form a cavity (void) in the adhesive layer, measures should be taken to prevent this, such as preheating after applying the adhesive or performing the bonding process under a reduced pressure.

5.4.3 Through-Silicon Via Module

For bumpless interconnects, including RDLs, the Damascene method is employed to simplify the processes. The Damascene wiring method also enables fabrication of high-density interconnects as well as Cu interconnects in the BEOL, as shown in Figure 5.6. For TSV processing, dry etching through the dielectric (device layer), Si, and adhesive layer is carried out. TSVs with a small aspect ratio, for example <3, have the advantage of shortening the process time for both etching and metal filling because those are proportional to the volume of the TSV, as shown in Figure 5.7. Assuming that the etching rate follows the mass transport limit reaction, the etching times, t and t_1, at different TSV diameters, D and D_1, and depths, d and d_1, follow $t_1/t = (D_1/D)^2 \times (d_1/d)$, that is, $t_1/t = 0.1$ at $D = D_1$, $d = 50$ μm, and $d_1 = 5$ μm, which suggests 1/10th of the etching time for the same TSV diameter and 1/10th of the depth. Reducing process time is the key for low-cost manufacturing. After TSV etching, a SiN or SiO_2 film is deposited by plasma-enhanced chemical vapor deposition (PECVD) to provide electrical insulation from the Si substrate. The barrier dielectrics at the bottom of the

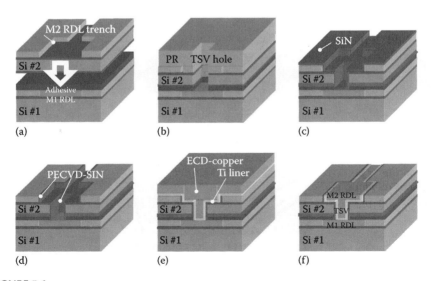

(a)

(b)

(c)

(d)

(e)

(f)

FIGURE 5.6
TSV formation and damascene Cu plug processes. After bonding of the thinned wafer to another wafer surface, RDL (redistribution line) patterning, TSV etching, Cu plug formation, and planarization by CMP are carried out. (a) Wafer level bonding, (b) TSV PR & Si DRIE, (c) PECVD, (d) SiN self-align etching, (e) PVD metal barrier and Copper ECD, and (f) planarization.

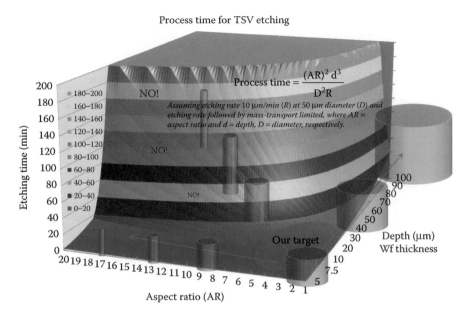

FIGURE 5.7
Silicon etching time versus aspect ratio of TSV and depth, assuming that the total etching volume follows the mass transport limit reaction. Etching time increases with increasing depth and aspect ratio.

TSV is removed by bias sputtering, and Ti/TiN and Cu are deposited on the barrier metal and the seed layer, respectively, by sputtering. For Cu plug interconnects, electrochemical-plated Cu (ECP-Cu) is used. ECP-Cu planarization is carried out by chemical–mechanical polishing (CMP), as shown in Figure 5.8. In case of permanent adhesive formed on Cu pad after CMP,

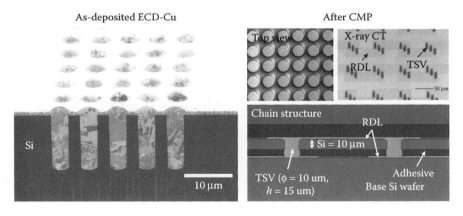

FIGURE 5.8
SEM images of Cu TSVs after ECD-Cu deposition and CMP. For ECD-Cu, to reduce polishing time in CMP, the overburden of Cu was controlled to be thin. Chain structure shows the cross-section of Cu pad and damascene TSV for two stack wafers.

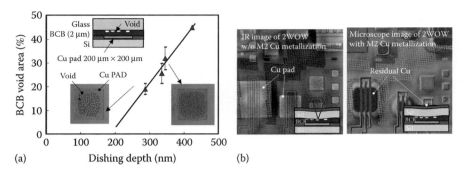

FIGURE 5.9
Occupied void area (BCB) formed on as a function of Cu pad as a function of dishing depth of Cu pad (a). Top view images of second wafer surface after Cu CMP using IR and optical microscope (b).

voids formation in adhesive occurs and depends on the dishing depth at Cu pad as Figure 5.9a. This dishing phenomenon (microsurface roughness) affects the next process in which Cu residue is removed by CMP in the second wafer process (Figure 5.9b). This suggests that Cu dishing have to reduce less than 200 nm in depth for the CMP process.

Figure 5.10 shows the via structure after etching and the leakage current, compared with the Bosch and direct etching methods [15–16].

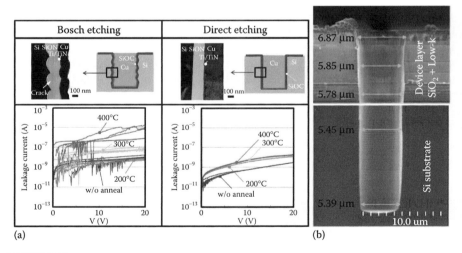

FIGURE 5.10
A comparison of leakage currents in two types of TSV samples made by Bosch etching and direct etching (a). Cracks are observed in the side walls of the TSV in the Bosch-etched sample, which had a rough interface called scalloping. The leakage current as a function of applied voltage after annealing at temperatures up to 400°C was measured. With increasing temperature, the leakage current increased but was two orders of magnitude higher in Bosch etching. By optimizing the scalloping shape, SEM images of TSV etched off through Cu/Low-k BEOL layer and Si are shown (b).

As Bosch etching was conducted by cyclic isotropic-etching and deposition, microsteps called scalloping were formed at the sidewall of the via. The scalloping causes cracks due to stress concentration in the dielectric layer and poor step coverage for thin films deposited by chemical vapor deposition (CVD) and physical vapor deposition (PVD). In contrast, anisotropic direct etching resulted in a smooth surface profile along the side wall. The leakage current in Bosch etching was one order of magnitude higher than that in direct etching. The leakage current was caused by Cu diffusion at the side wall of the TSV that took place at a thinner part of the dielectric containing cracks. Thus, direct etching of TSVs, in other words etching that is capable of producing a smooth sidewall, is required for TSV processing. In addition, low-aspect ratio Cu TSVs cause lower stress concentration in thermal processing compared with high-aspect ratio Cu TSVs. Low stress reduces Cu deformation and stress propagation to the device regions.

5.4.4 Packaging Module (Singulation/Packaging Module)

After wafer stacking, the same procedure as in conventional packaging (bumping, singulation by dicing, resin packaging) is followed. For dicing of a seven-level wafer stack, the adhesive layer and silicon chips were found to be free of defects or delamination, as shown in Figure 5.11. After the stacked chips were packaged with epoxy resin, they were subjected to heat stress testing at temperatures of −65°C to 150°C. Scanning acoustic tomography (SAT) was adopted for internal observation, and after up to 100 repeated heat stress tests, no delamination was found at the interfaces between the molding compound and chips, nor at the chip stack interfaces, as shown in Figure 5.12.

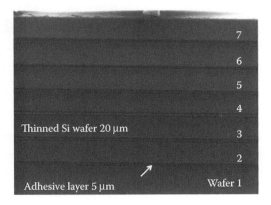

FIGURE 5.11
SEM cross-sectional image of seven-wafer stack after dicing. No crack and delamination of bonding material were found.

Thermal stress test (−65°C~150°C, 10 min/10 min)

Pristine	1 cycle	10 cycle	100 cycle

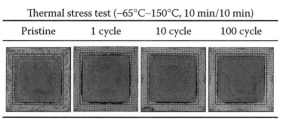

Scanning acoustic tomography image

FIGURE 5.12
Thermal stress testing at temperatures of −65°C to 150°C for 100 cycles, observed by scanning acoustic tomography (SAT).

5.5 Device Characteristics after Ultrathinning

5.5.1 Retention Time Change of Dynamic Random-Access Memory after Thinning

The effect of wafer thinning on device performance was examined to evaluate the thinning limits for 300 mm silicon wafers, as shown in Figure 5.13 [10,11,17]. The ferroelectric RAM (180 nm-node FRAM or FeRAM) device wafer was thinned down to 9 μm. There is a capacitor formed of

Device Node	FRAM (VLSI2010) ~180 nm	SRAM (IEDM2009) 45 nm (Lg 35 nm)	DRAM (IEDM2014) 40 nm (2 Gb)
Wafer thickness	9 μm	7 μm	4 μm
SEM picture			
Electrical property			

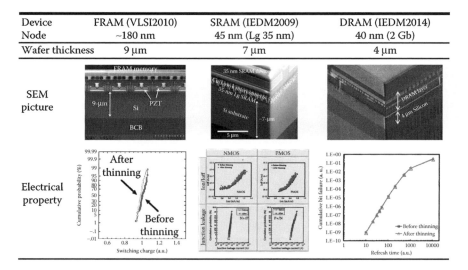

FIGURE 5.13
Device characteristics before and after wafer thinning for ferroelectric random access memory (FeRAM), MPU (SRAM), and DRAM. In the FeRAM, the quantity of residual polarization charge (Qsw) was evaluated; in the SRAM, the on- and off-current, and leakage current; and in the DRAM, the retention time. There were no significant changes in these characteristics after wafer thinning down to 9 and 4 μm.

lead-zirconate-titanate PbZrxTi1-xO3 (PZT) dielectric underneath the bit line (M1). There was no significant shift of switching charge within the range of the design specifications. In the case of a microprocessor (45 nm-node MPU) device, thinning was carried out down to 7 μm. For both n-MOSFETs and p-MOSFETs, the on-current (Ion), off-current (Ioff), and roll-off characteristics of threshold voltage (Vth) were nearly the same as those before thinning. In the case of DRAM devices, which are most sensitive to induced leakage current due to defects, the wafer was thinned to a final thickness of 4 μm, which is about 0.5% of the thickness of the bulk wafer (725 μm) and thinner than the device layer. The TTV within the 300 mm wafer was low enough to realize multistacking: A TTV of 1.02 μm was achieved at an average thickness of 4 μm. The wafer was annealed during the bonding process at around 250°C for 60 minutes. No significant change with retention time (refresh time) in the entire wafer before and after thinning was observed for the Si thicknesses of 40, 20, 8, and 4 μm [10]. This suggests that the thinning process developed in this study did not affect the junction leakage current, which degrades the retention time more sensitively than other leakage phenomena such as subthreshold leakage, capacitor dielectric leakage, and gate-induced drain leakage (GIDL).

5.5.2 Ultrathinning and Estimation of Critical Thickness

Wafer thinning was carried out as follows: coarse grinding (#320 grit size) down to ~50 μm, fine grinding (#2000 grit size) to <20 μm, and post-thinning using CMP as an option. Figure 5.14 shows a schematic of the thickness

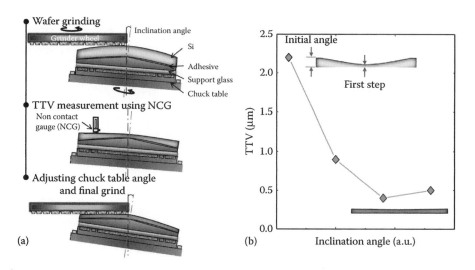

FIGURE 5.14
Improvement in total thickness variation (TTV) using so-called auto-TTV process employing NCG (noncontact gauge) methods (a) and TTVs for 300 mm wafer with various inclination angles (b).

variation control in the grinding process. The wafer thickness uniformity after grinding was determined by the contact angle between the wheel and wafer surface [18]. Usually, the wafer was very slightly bowed after bonding due to elastic deformation of the temporary adhesive at the wafer edge, and this shape was also reflected in the contact angle. By adjusting the contact angle to follow the geometric shape of the wafer, the TTV decreased to as low as 0.5 μm within the 300 mm wafer. With these thinning processes, the thickness of the damaged layer, including point defects such as vacancy-type defects, was decreased from micrometer level to several nanometers when evaluated by transmission electron microscopy (TEM) and positron annihilation spectroscopy analyses, as shown in Figure 5.15 [19,20]. To evaluate the critical thickness, thinning was carried out on a 300 mm DRAM wafer (DDR3 SDRAM) down to ~2 μm, and Cu was intentionally formed at a density on the order of 10^{13} atoms/cm^2 on the back side after thinning. Then, the wafer was annealed at 250°C for 60 minutes. Retention degradation and yield loss within the wafer were observed at thicknesses less than 2.6 μm, as shown in Figure 5.16 [10]. According to those results, in case of a DRAM, there was a critical thickness of the Si wafer of around 2 μm.

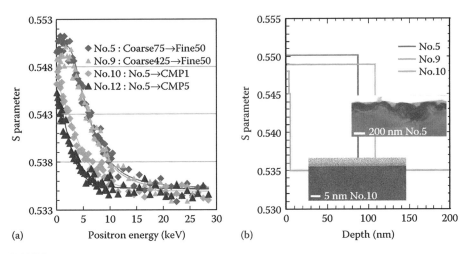

FIGURE 5.15
Doppler broadening spectrum of positron annihilation spectroscopy analyses for wafer backside after thinning. (a) S parameter as a function of incident positron energy E for fine grinding (Samples No. 5 and 9) and CMP (Samples No. 10 and 12). The number in terms such as "Coarse75" represents 75 μm coarse grinding. (b) Depth profiles of S parameter for fine grinding and CMP. The S parameter is normalized by that of Sample No. 12, which is assumed as a reference. Thicknesses of the defect layer were ~200 and 3–4 nm for fine grinding and CMP, respectively, observed by TEM analysis, which agrees with the depth profiles.

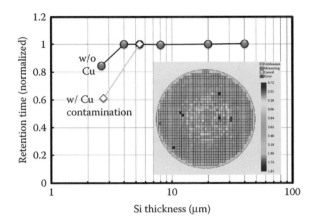

FIGURE 5.16
Retention time of DRAM as a function of Si thickness. Cu was formed at a density on the order of 10^{13} atoms/cm^2 at the back side after thinning. Then wafers were annealed at 250°C for 60 minutes. Normalized retention time was the time at 80% yield. Wafer map represents Si thickness, where the average thickness and TTV were 2.66 and 1.6 μm, respectively.

5.5.3 Cu-Diffusion Phenomenon at Ground Surface

A Cu-diffusion phenomenon was evaluated at the back side of the ground Si surface [19]. Island-like Cu aggregations were observed from back surface, as shown in Figure 5.17, in which Cu was detected with a high-angle annular dark field (HAADF) scanning TEM and energy-dispersive X-ray spectroscopy (EDX). The sample was annealed at 250°C for 60 minutes. In order to

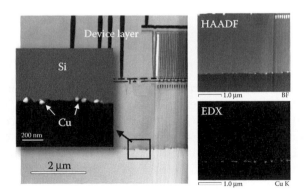

FIGURE 5.17
Cross-sectional TEM image of DRAM wafer after grinding and Cu contaminated. Cu aggregation at the back side of the Si surface was observed by HAADF-STEM and EDX analyses.

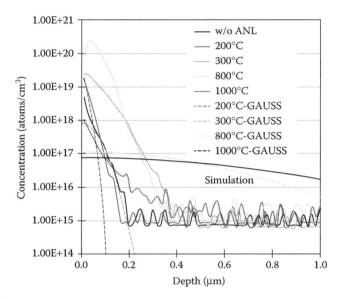

FIGURE 5.18
Backside SIMS depth profiles of Cu in Si after back-side grinding for various annealing temperatures, and Gaussian distribution of Cu into Si (dotted line) calculated using the diffusion kinetic model.

compare Cu diffusions, the annealing temperature was gradually increased to 1000°C in 30 minutes under the vacuum during the grinding process, and Cu was formed at the back side as well. The depth profile of the Cu was measured by time-of-flight secondary ion mass spectrometry (TOF-SIMS), as shown in Figure 5.18. The diffusion profile based on a diffusion kinetic model, the dotted line in Figure 5.18, followed the equation:

$$C = \frac{Q}{\sqrt{\pi D_0}} e^{\left(-\frac{x^2}{4Dt}\right)}$$

C and Q were Cu concentration in depth and Cu concentration at Si surface, respectively, where the diffusion coefficient of Cu in Si, $D_0 = 4.7 \times 10^3 \times \exp(-0.43 \text{ eV/kT})$ cm²s⁻¹ was used [21]. According to the SIMS results, the diffusion depth of Cu was around 400 nm even at 800°C, whereas Cu is predicted to diffuse by more than 1 μm from the diffusion model. This suggests that the Cu was stable at the damaged layer at the back side of the wafer and thought to be damaged layer having point defects that act as gettering sites for Cu atoms. Recrystallization of silicon occurred at a high temperature (>800°C). As those damaged layers consisted of atomic level vacancy-type defects, these vacancy defects were substituted by Cu atoms, which led to a stable state preventing further Cu diffusion.

5.6 Characteristics of Low-Aspect Ratio Through-Silicon Via Interconnects

5.6.1 Step Coverage and Cu Diffusion

For high-reliability Cu metallization, formation of a conformal barrier layer is necessary for preventing Cu diffusion into the Si substrate [15,22]. Figure 5.19 shows a cross-sectional TEM image of TSVs at the side wall of a 10 μm diameter via. The aspect ratio was about 1.5. The dielectric barrier layer of the SiON film deposited by the conventional PECVD process had a thickness at the bottom of the via that was 30% of that at the top of the via. A very thin barrier metal film composed of Ti/TiN deposited by the PVD process has no film continuity because of the low coverage. Even with a low-aspect ratio TSV, cracks occurred at the scallop-shaped microsteps due to the Bosch process, whereas no cracks were observed in the smooth side wall processed by anisotropic direct etching. These cracks cause leakage current. Cu diffusion increases when the density of the SiON layer decreases. In the case of a SiON layer with a relative density of 0.6 deposited at 100°C, Cu easily reached the interface between the dielectric layer and the Si substrate. Figure 5.20 shows the critical thickness (defined as the thickness that prevents Cu diffusion within the dielectric layer) of dielectric films as a function of the relative density. Leakage current occurred at sidewall of TSV due to Cu diffusion has been observed by the infrared-optical beam irradiation (IR-OBIRCH) method [23]. The point at a relative density of 0.8 illustrates the case for the cap layer of Cu interconnects used in Cu/low-k studies at a comparatively higher deposition temperature. For a relative density below 0.5, a dielectric layer thicker than 1 μm would be needed to prevent Cu diffusion

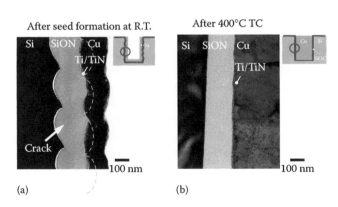

FIGURE 5.19
TEM cross-sectional image of TSV after SiON/Ti/TiN/Cu formation. Scallop shape at the side wall and crack propagation at the SiON layer were found for Bosch etching (a). No crack was observed for direct etching, even after annealing (b).

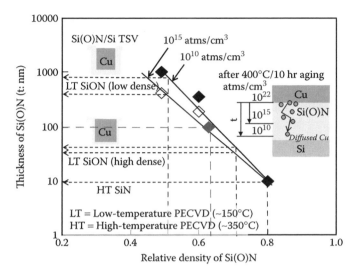

FIGURE 5.20
Critical thickness for Cu barrier of dielectric layer as a function of film density. Density represents relative value with respect to that of bulk Si_3N_4 (3.44 g/cm³).

into the Si substrate. This suggests that the effective Cu area decreases with an increase in the thickness of the dielectric, which limits the scalability of the Cu TSV diameter, for example, a Cu TSV diameter of only 1 μm remains at a 3 μm thick dielectric.

5.6.2 Stresses in Cu Through-Silicon Vias

For the bumps attached to the device layer, delamination occurs at the Cu/low-k interconnect, as shown in Figure 5.21a. This is due to excessive stress induced during the process of bonding process the low-mechanical stress Cu/Low-k interconnects. Furthermore, the bumps themselves tend to fracture under the high-stress concentration due to a large CTE mismatch between the bumps and underfill materials. As bumps are absent in bumpless TSVs, no delamination is observed, as shown in Figure 5.21b. The stress induced at the Cu/low-k interconnect layer was estimated by finite element method (FEM) analysis using Chebyshev's orthogonal function obtained by the design of experiment (DOE), and the ratio of the thickness of the adhesive layer ($T_{Adhesive}$) to the Si (T_{Si}) as a function of the aspect ratio of the TSV is shown in Figure 5.22. These results suggest that the induced stress increases when the thickness of the adhesive layer and aspect ratio increases. Therefore, the use of small TSVs and a thin adhesive layer, in other words, a no bump structure, provides a better solution for stress reduction.

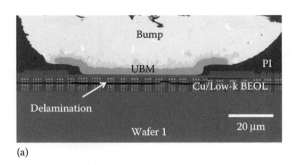

(a)

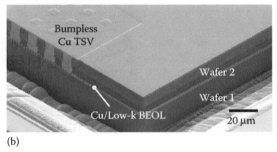

(b)

FIGURE 5.21
Cross-sectional SEM images of (a) bump and (b) bumpless TSV formed on Cu/Low-k BEOL interconnects.

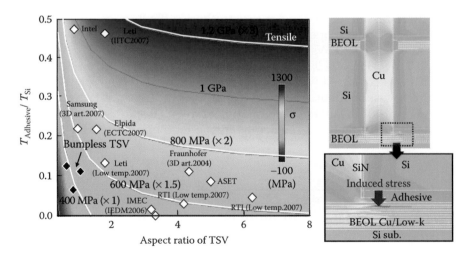

FIGURE 5.22
Stress at Cu/Low-k interconnects layer versus the ratio of thickness of adhesive layer to thickness of Si as a function of aspect ratio of TSV.

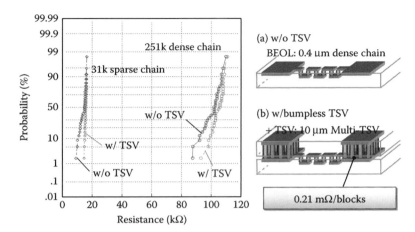

FIGURE 5.23
Via chain resistance cumulative failure distribution of Cu BEOL interconnects with and without Cu-TSVs. Schematic of test structure for electrical measurement: (a) 65-nm Cu interconnects and (b) with multiple TSVs in two-wafer stack.

5.6.3 Electrical Characteristics

Figure 5.23 shows the cumulative failure distribution of the Cu-BEOL via chain resistance and with multiple bumpless Cu TSVs [22]. The Cu-BEOL via chains having a total number of vias equal to 31×10^3 (sparse chain) and 251×10^3 (dense chain) were compared. The diameter of the TSV and total depth of the TSV, including the adhesive layer, are 5 and 15 µm, respectively. The contact resistance of the Cu-TSV to the Cu-BEOL is estimated to be 0.21 mΩ. There are no open failures or significant changes in the resistance distribution with the TSVs. The leakage current between TSVs blocks was as low as 2.3×10^{-11} A at 4.0 V. Thermal stress testing of a structure in which a Cu-TSV was contacted with a BEOL chain was carried out from −55°C to 125°C. In the case of a dense chain, the initial resistance of 104 kΩ was changed to 100 kΩ and 101 kΩ after TSV formation and 1000 test cycles, respectively. This suggests that there was no degradation after thermal stressing.

5.7 Thermal Resistance of Bumpless Interconnects and Thin Wafers

Heat dissipation becomes more difficult in a 3DI structure due to the thermal resistance of the stack of multiple devices and interconnect-dielectrics layers. The total thermal resistance for a stack of multiple devices was

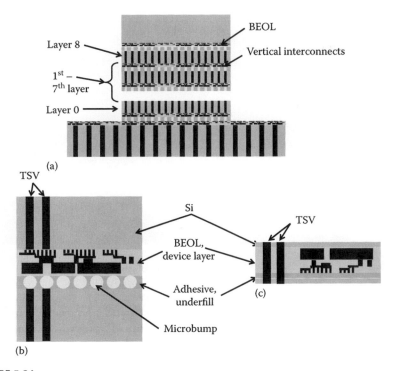

FIGURE 5.24
Schematic diagrams of thermal resistance estimation for Si substrate having a BEOL layer and TSV vertical interconnections with and without bumps: (a) nine-device stacked structure with bumps, (b) details of TSV and bump with underfill, and (c) TSV (no bump) and ultrathin wafer with organic bonding adhesive.

evaluated using bump and bumpless interconnects [24]. The thermal resistance of a Si substrate having a BEOL layer and TSV vertical interconnects with and without bumps was estimated, as illustrated in Figure 5.24. There are polymer and composite materials, such as underfill in the case of the bump interconnects and bonding adhesives in the case of bumpless interconnects. The thermal resistance *R*th of the vertical interconnects and the total thermal resistance were calculated using the FEM and a thermal network method, respectively. Assuming a stack of eight DRAMs and one controller wafer, the thermal resistance was estimated by the following sequence: estimating the effective thermal conductivity of each layer and calculating the temperature rise using the thermal network method. The thermal conductivities used were 148, 160.5, and 1.44 W/mK for Si, Si with TSVs, and BEOL, respectively [25]. Figure 5.25 shows thermal resistance as a function of normalized via occupancy. The thermal resistance decreased as both the thicknesses of the underfill material and bonding

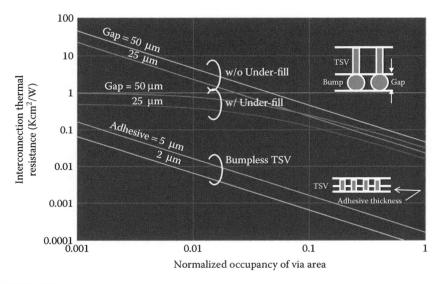

FIGURE 5.25
Thermal resistance of interconnects as a function of normalized occupancy of via area, where microbump and TSV (no bump) are compared.

adhesive decreased, and two orders of magnitude lower the bumpless structure when compared with the conventional bump structure. This is due to thickness reduction of the polymer layer and bumpless structure, which has higher thermal resistance and the shortened interconnects. This suggests that only 1% the total metal area of bumpless TSVs is required to achieve a thermal resistance equivalent to that of the bump structure. The total thermal resistances of a stack of eight wafers with and without bumps are estimated as shown in Figure 5.26. The bump size and pitch were 25 μm (thickness of underfill material) and 50 μm. For bumpless TSVs, the TSV size, pitch, and thickness of the bonding adhesive were 10, 20, and 5 μm, respectively. The total thermal resistances were 1.54 and 0.46 Kcm^2/W for the bump and bumpless structures, respectively. According to these total thermal resistance values, the temperature increase in the wafer stack was estimated as shown in Figure 5.27, where the temperature indicates the relative increase in temperature from the bottom wafer. The temperature rises with an increase in the number of stacked wafers and becomes approximately 20°C for the bump structure, which is about four times higher than that of the bumpless structure. Therefore, if such a temperature increase is allowed, a stack containing five-times more wafers will be possible with bumpless TSV interconnects.

		Components	Equivalent thermal conductivity (W/mK)	Microbump 51.4 µm pitch 25 µm bump		Bumpless 512*16 TSV 5 µm gap	
				Thickness (µm)	Thermal resistance (Kcm²/W)	Thickness (µm)	Thermal resistance (Kcm²/W)
DRAM	Top Chip Rth1	Si	148	150	0.049	5	0.00034
		BEOL	1.44	15	0.104		
			3.99			15	0.038
		Interconnection microbump	2.54	20	0.079		
		Interconnection bumpless	2.56			5	0.020
	2–8 Chip 7 layers Rth2–8	Si with TSV	160.5	50	0.003	5	0.00031
		BEOL	1.44	15	0.104		
			3.99			15	0.038
		Interconnection microbump	2.54	20	0.079		
		Interconnection bumpless	2.56			5	0.020
Logic RthL		Si with TSV	160.5	50	0.003	5	0.0003
		Total thermal resistance	Rth1+7*Rth2–8+RthL		1.54		0.46

FIGURE 5.26
Summary of total thermal resistance of TSV + microbump and TSV (no bump) structure.

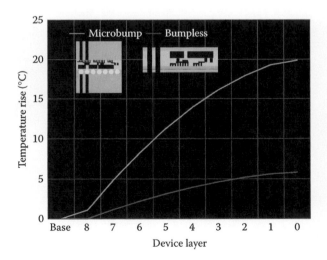

FIGURE 5.27
Temperature rise for each layer in stacked devices in bump and no-bump (bumpless) structures.

5.8 Concurrent Manufacturing Using WOW

5.8.1 Enhancement of Memory Capacity

As the WOW process, *thinning-first, back-to-face* bonding, and then *TSV-last*, allows thinning of silicon wafers down to 4 μm without any degradation of the device characteristics, the total wafer thickness, including the device layer and the adhesive layer, becomes 10–20 μm, which is 1/10th to 1/100th of the thickness of conventional bump interconnects using TSVs. Even if the number of stacked wafers is 100, assuming that the wafer thickness is 10 μm, the total thickness after stacking is only 1 mm. If a cubic space can be utilized, 1000 device layers can be stacked. By following these multilevel stacking processes, with a memory device such as a DRAM fabricated with 22 nm technology and having a memory density of 30 GB/cm², when four, eight, sixteen, and so on, of these devices are stacked, the total capacity of the memory device can be increased to 120, 240, 480 GB, and so on, respectively. In case of Flash nonvolatile memory, having 3D-transitor memory cells, terabit-capacity memory can be realized by stacking only 4 or 8 wafers. In contrast, to achieve equivalent capacity with a single wafer using extreme scaling or 3D transistors, 1 nm processing technology or more than 128 transistor stacks will be required, as shown in Figure 5.28.

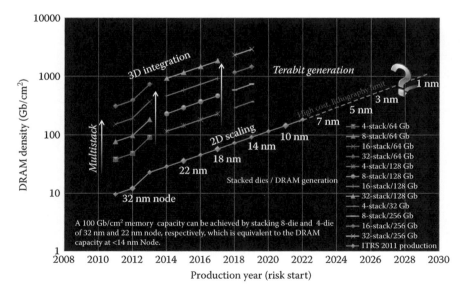

FIGURE 5.28
Trend of DRAM density using 2D conventional scaling and 3D multistacking using existing DRAM. DRAM capacity in the 3D case corresponds to the number of stacked dies, assuming that redundancy is eliminated by cell blocks at each layer.

5.8.2 Yield in Wafer Stacking

In the case of die-to-die series connections, the total yield in wafer stacking can be estimated by Y^n, where Y and n are the nondefective die yield of one wafer and the number of stacked wafers, respectively, as shown in Figure 5.29 [26]. The total yield of stacked wafers without any design improvements decreases with increasing number of stacked wafers, for example, $Y = 0.9$ and four stacked wafers give a total yield of $Y^4 = 0.656$.

By using multiple channel or memory mat for designing the architecture of the memory layout, the defect area will be limited at the channel region. Thus, the memory capacity within one memory die, excluding failed channels, is maintained. In other words, when one memory die has multiple channels, the loss of memory capacity would be reduced even at the same defect density. As the bumpless TSVs and ultrathin WOW process enable interconnections with high density at the minimum physical length, those channels are vertically connected to the controller independently, as shown in Figure 5.30. For instance, if one channel requires 64-bit I/Os, the total vertical interconnects from 4 dies correspond to a total of 4096 bit I/Os. This is difficult to achieve with small TSV area using conventional bump processes because the density of bumps and TSV processes is limited to the bump size and pitch (~50 μm).

In the case of a conventional die-level stack, the yields of four good dies and three good dies for the case where four wafers are stacked, based on the combination probability at $Y = 0.68$, are 0.17 and 0.86, respectively [3,4,25]. With a memory density of 30 Gbit/cm^2 (equivalent to 22 nm technology), the memory capacity of four dies and three dies reaches 120 and 90 Gbit, respectively. If the number of effective chips is 1700 per 300 mm wafer at a chip size of ~0.4 cm^2 and 8 Gbit/die, the expected number of chip sets per unit memory capacity is 289/32 Gbit (4 stacked dies), 1462/24 Gbit (3 stacked dies), and 1088/8 Gbit (single die). As the multichannel design supports further

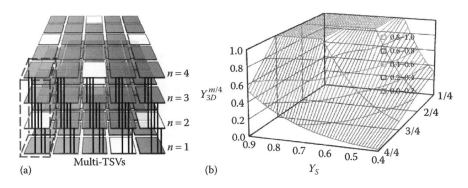

FIGURE 5.29
Schematic of stack wafers of a combination in four-level stacked wafer (a), and yield for four-level 3D wafer stack and a comparison of good-die combination at die size = 1.148 cm^2 and defect density $D_0 = 0.2/\text{cm}^2$ (b).

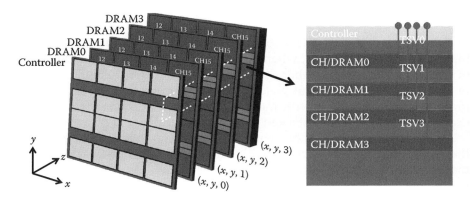

FIGURE 5.30
Schematic diagram of DRAM stack structure. Here, one memory die has 16 channels (CH0 to CH15) in total. Wafers following DRAM2, DRAM1, DRAM0, and controller are stacked to DRAM3 (base wafer) using bumpless interconnects and a WOW process. Bumpless interconnects are connected independently to the controller die from each channel of the DRAM layers.

yield improvements as described earlier, sufficiently high memory capacity with high yield can be expected in the WOW memory stack.

5.8.3 Benefits of Co-Engineering

Achieving a capacity of over 24 Gbit per single chip would require technology two or three generations ahead, such as 10–14 nm technologies. Defects, however, will become a bottleneck due to the difficulty in controlling defects. For instance, subnanometer roughness control of the gate width along the gate line must be achieved after patterning. This is because, as scaling proceeds, the so-called stealth defects (invisible defects) increase proportionally, and the control of process variations approaches its limit. When stealth defects become dominant, variations cannot be improved statistically and thus die yield deceases with scaling. Considering the technology roadmap, the issues of scaling technology and technology for fabricating three-dimensional structures are often discussed separately; for example, scaling belongs to the front-end wafer process, whereas 3DI belongs to back-end packaging. However, these two technologies are not always mutually exclusive. Scaling would be relieved of the stringent requirements by co-engineering using three-dimensional WOW technology combined with mass-production technology. In other words, a sufficiently long learning period would be ensured, and further cost reductions could be expected by concentrating on the control of variations among generations and shortening the process. As cost reduction requires the adoption of advanced lithography technology, advanced lithography and peripheral support facilities account for one-third to one-fourth of the total cost of a manufacturing line. In short, although useful for reducing chip cost, scaling is very burdensome in terms of capital investment, and the chip cost per unit area is currently saturated, even after taking account of shrinking chip sizes. In combination with three-dimensional stacking to overcome such problems, a roadmap toward high-density integration backed up by production costs can be made. This is because the capital investment for 3D wafer processes is not high compared to that of lithography. Moreover, keeping the wafer shape as-is for stacking ensures compatibility with manufacturing facilities in front-end processing and helps utilize the mature process technologies developed for wafer processing. If the processes up to three-dimensional stacking can be handled as units in the manufacturing line, the throughput will be 1/100th of that in stacking starting with chips. Therefore, future semiconductor manufacturing is expected to advance with a roadmap in which the number of stacked wafers, the wafer thickness, and the number of TSV interconnects serve as indices, as shown in Figure 5.31.

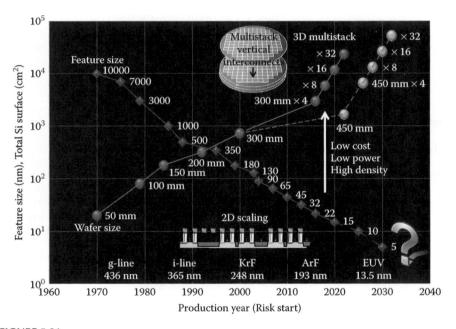

FIGURE 5.31

Trends of two-dimensional (2D) scaling and wafer size including total Si surface of wafer stack. Conventional scaling will face difficulties such as physical limits and inability to minimize costs, whereas 3D integration will become superior to scaling. By combining conventional two-dimensional integration (2DI) with three-dimensional stacking to overcome such problems associated with device scaling and increasing wafer size, it is possible to make a roadmap toward high-density integration backed up by production costs. In volume production, 3D wafer stacking (WOW) enables a lower cost than chip-on-chip (COC) and high-density integration, reaching Terabit level. Bumpless interconnects using TSVs and ultrathinning provide high-density I/Os connecting top and bottom device layers and achieve a small form factor 1/10th that of bump structures.

5.9 Summary and Conclusions

WOW technology and bumpless interconnects using TSVs for 3D stacking in wafer form have been described. It was found that an optimized wafer-thinning process for determining the stack thickness does not cause degradation of the device characteristics in advanced commercial devices, even with the smallest thickness of 4 μm. As bumpless interconnects using TSVs can be connected directly to the upper and lower substrates by self-alignment, the package thickness can be reduced by an amount equivalent to the thickness of electrodes, such as bumps, which are not necessary when bumpless interconnects are used in combination with wafer thinning. As the design pitch of TSVs is determined by the bump size, high-density TSVs can be formed in bumpless interconnects by following TSV-patterning processes. At the same time, size reduction of the finished shape allows the wiring between the upper and lower chips to be made shorter, which reduces the total wiring impedance and makes it easier to ensure high bandwidth with higher energy efficiency. Furthermore, by stacking wafers, high-density integration and system block arrangements become more flexible, and the design space is extended.

References

1. Cisco, 2013. http://www.cisco.com/c/dam/en_us/about/business-insights/docs/ioe-value-at-stake-public-sector-analysis-faq.pdf, By comparison, the "Internet of Things" (IoT) refers simply to the networked connection of physical objects (doesn't include the "people" and "process" components of IoE). IoT is a single technology transition, while IoE comprises many technology transitions (including IoT).
2. T. Ohba, N. Maeda, H. Kitada, K. Fujimoto, K. Suzuki, T. Nakamura, A. Kawai, and K. Arai, *Microelectron. Eng.*, **87**: 485–490, 2010.
3. T. Ohba, *Electrochem. Soc. Trans.*, **34** (1): 1011–1016, 2011.
4. T. Ohba, Y. S. Kim, Y. Mizushima, N. Maeda, K. Fujimoto, and S. Kodama, *IEICE Electron. Expr.*, **12** (7): 1–14, 2015.
5. G. Moore, *Electron. Mag.*, **38** (8), April 19, 1965.
6. J. U. Knickerebocker, P. S. Andry, B. Dang, P. R. Horton, M. J. Interrante, C. S. Patel et al., *IBM J. Res. Dev.*, **50** (4/5): 553–567, 2006.
7. M. Koyanagi, T. Nakamura, Y. Yamada, H. Kikuchi, T. Fukushima, T. Tanaka, and H. Kurino, *IEEE Trans. Electron Devices*, **53** (11): 2799–2808, 2006.
8. E. Beyne, P. D. Moor, W. Ruythooren, R. Labie, A. Jourdain, H. Tilmans, D. S. Tezcan, P. Soussan, B. Swinnen, and R. Cartuyvels, *IEEE IEDM Technical Digest*, pp. 495–498, 2008.
9. H. Kitada, N. Maeda, K. Fujimoto, K. Suzuki, A. Kawai, K. Arai, T. Suzuki, T. Nakamura, and T. Ohba, *IEEE IITC*, pp. 107–109, 2009.

10. Y. S. Kim, S. Kodama, Y. Mizushima, T. Nakamura, N. Maeda, K. Fujimoto, A. Kawai, K. Arai, and T. Ohba, *IEEE IEDM Technical Digest*, pp. 189–192, 2015.
11. Y. S. Kim, S. Kodama, Y. Mizushima, N. Maeda, H. Kitada, K. Fujimoto, T. Nakamura et al., *IEEE VLSI Symposium*, pp. 22–23, 2014.
12. P. S. Foster, E. Ecker, E. Rutter Jr., and E. S. Moyer, U.S. Patent 5,882,836, 1999.
13. Y. H. So, D. M. Scheck, C. L. Murlick, G. S. Becker, and E. S. Moyer, World Patent 9631805, 1996.
14. Y. Kwona, A. Jindala, R. Augurb, J. Seokc, T. S. Calea, R. J. Gutmanna, and Jian-Qiang, *J. Electrochem. Soc.*, **155** (5): 280–286, 2008.
15. H. Kitada, N. Maeda, K. Fujimoto, Y. Mizushima, Y. Nakata, T. Nakamura, and T. Ohba, *Jpn. J. Appl. Phys.*, **50** (5): 05ED02, 2011.
16. D. Diehl, H. Kitada, N. Maeda, K. Fujimoto, S. Ramaswami, K. Sirajuddin, R. Yalamanchili et al., *Microelectronic Eng.*, **92**: 3–8, 2011.
17. N. Maeda, Y. S. Kim, Y. Hikosaka, T. Eshita, H. Kitada, K. Fujimoto, Y. Mizushima et al., *IEEE VLSI Symposium*, pp. 105–106, 2010.
18. Y. S. Kim, N. Maeda, H. Kitada, K. Fujimoto, S. Kodama, A. Kawai, K. Arai, K. Suzuki, T. Nakamura, and T. Ohba, *Microelectron. Eng.*, **107**: 65–71, 2013.
19. Y. Mizushima, Y. S. Kim, T. Nakamura, R. Sugie, H. Hashimoto, A. Uedono, and T. Ohba, *J. Appl. Phys.*, **53**: 05GE04, 2014.
20. A. Uedono, Y. Mizushima, Y. Kim, T. Nakamura, T. Ohba, N. Yoshihara, N. Oshima, and R. Suzuki, *J. Appl. Phys.*, **116**: 134501-1–134501-5, 2014.
21. E. R. Weber, *Appl. Phys.*, **30** (1): 1–22, 1983.
22. H. Kitada, N. Maeda, K. Fujimoto, Y. Mizushima, Y. Nakata, T. Nakamura, W. Lee, Y.-S. Kwon, and T. Ohba, *IEEE IITC*, 2010.
23. Y. Mizushima, H. Kitada, K. Koshikawa, S. Suzuki, T. Nakamura, and T. Ohba, *J. Appl. Phys.*, **51**: 05EE03, 2012.
24. H. Ryoson, K. Fujimoto, and T. Ohba, *JIEP Proceedings of the International Conference on Electronics Packaging* (ICEP), pp. 522–525, 2017.
25. K. Matsumoto, S. Ibaraki, K. Sueoka, K. Sakuma, H. Kikuchi, Y Orii, F. Yamada, K. Fujihara, J. Takamatsu, and K. Kondo, *29th IEEE SEMI-THERM Symposium*, 2013.
26. T. Ohba, *IEEE Proceedings of IITC*, 12.1, 2010. doi:10.1109/IITC.2010.5510747.

6

Three-Dimensional Integration Stacking Technologies for High-Volume Manufacturing by Use of Wafer-Level Oxide-Bonding Integration

Spyridon Skordas, Katsuyuki Sakuma, Kevin Winstel, and Chandrasekharan Kothandaraman

CONTENTS

6.1 Introduction

It is a very well understood fact that device and interconnect (IC) scaling in semiconductor technology applications is becoming increasingly difficult and more expensive with each new technology node. The traditional model of continuous electric field scaling that had been for decades the key to the success and growth of the microelectronics industry [1], and which enabled Moore's law, has ceased to be applicable, at least not in a straightforward way. However, the semiconductor industry is still expected to follow Moore's law in the economic sense and continues to deliver significant improvements in device and chip performance to enable yet more applications, with each new generation of technology, while still supporting concomitant decreases in cost per function. The driving force for the last two to three decades is the incessant implementation of new innovations to enable the continuation or even acceleration of the business model based on Moore's law. Many new materials have been introduced along with new advanced processes and increasingly complex integration schemes and device architectures. This has been undoubtedly necessary as the traditional silicon-based materials and traditional complementary metal–oxide–semiconductor (CMOS) architectures reached their physical limitations as the scaling trends demanded structures increasingly in the nanoscale. Furthermore, interconnect scaling has also resulted in resistance–capacitance (RC) delays now becoming a major contributing factor to limiting overall performance of systems, which were usually significantly gated primarily by transistor performance in the past. All these innovative approaches have driven collectively the tooling and fab costs much higher and have increased the risks, cost, complexity, and the time involved to deliver technologies through the pipeline, from early research and development (R&D) to manufacturing ramp up, and to fully mature product. To make matters worse, the performance gains with each technology node are not as sizeable or straightforward as in the era of traditional scaling, which makes the cost-benefit considerations for future investments even more challenging. It is becoming increasingly apparent that revolutionary rather than evolutionary solutions are needed to sustain or even accelerate again the desired rate of combined improvements in the performance and the costs of the technology. This continued demonstration of technological prowess is still the hallmark of the semiconductor industry and, combined with the cost efficiencies, it is expected by both enterprise and individual customers and has been an essential factor in the proliferation of semiconductor technology applications across all areas of human life and enterprise.

6.1.1 Rationale for Three-Dimensional Integration

In keeping with this spirit of urgent and revolutionary innovation, one of the most promising avenues toward the desired direction of increased performance and cost-effective system scaling is three-dimensional integration (3DI) technology [2,3]. According to the 3DI technology paradigm, system performance scaling can be achieved by the stacking of chips in a homogeneous or heterogeneous fashion that can enable performance and functionality gains even beyond Moore's law. Due to its use of the third dimension, 3DI can realize advantages with regard to form factors and can achieve higher device density per unit area compared to two-dimensional (2D) architectures [4]. Furthermore, the stacking of chips can enable significant reduction of length for long-range interconnect paths relative to traditional 2D designs, which can in turn enhance performance and also can contribute to lower power demands than the 2D designs [5]. This 3D stacking of chips can be realized in a variety of ways, which include package-to-package stacking, die-to-die stacking, die-to-wafer stacking, and wafer-to-wafer stacking. The main advantage of the package-to-package and die-to-die stacking is that they inherently require and enable the selection of known good die (KGD) for the stacking, thus providing guaranteed higher yield versus other stacking schemes [5,6]. However, the performance improvements compared to traditional 2D designs can be limited due to the nature of the bonding technology and integration schemes. In addition, these involve a higher cost penalty because typically each die needs to be individually handled during the built of the stack. In these 3D integration schemes, the through-silicon vias (TSVs) that are necessary for chip-to-chip interconnections in the package and die-level stacks are typically of relatively large size. This is definitely a limiting factor with regard to the overall bandwidth and performance of these systems, as they limit the chip-to-chip interconnection density. On the other hand, 3D stacking by employing wafer-to-wafer bonding can allow for the use of traditional silicon fab wafer-level processing, which in turn can enable the scaling of TSVs by a factor of >20 [7,8]. This in turn enables much better performance and higher bandwidth with lower associated manufacturing cost, as the stacking of wafers can enable the massively parallel stacking of chips and inherently would be more suitable and efficient for high-volume manufacturing. However, wafer-stacking technology itself has its own challenges, primarily due to the compounded yield losses that can occur due to any nonfunctional die in the wafers used as these stacks are manufactured. It would therefore necessitate very high wafer-level yields or the implementation of efficient yield–loss mitigation strategies at the design level through fault-tolerant circuit design and the integration in the design of repair circuit architectures. These would enable the mitigation of potential yield losses but would also increase the design complexity with respect to chip real estate. Also, wafer-level integration places strict requirements for high process yield for the wafer bonding and thinning modules. Overall, wafer-level bonding is very promising for the realization of cost-effective, high-volume manufacturing of high-performance 3DI stacks. With the ever-increasing consumer demand for higher performance chips/stacks of smaller area size, increased device density, and reasonable cost, which is driven primarily by the memory business and the backside image sensor business, the industry is increasingly considering wafer-level stacked technology. It is noteworthy that the scaling of TSVs depends on the aspect ratio of TSV diameter to the thickness of thinned silicon wafer. It follows that the thinned silicon wafer thickness needs to scale with the TSV diameter so as to achieve a fixed aspect ratio as much as possible, and therefore all the related enabling processes (bonding, thinning, etching, TSV liner deposition, and metal fill) must be able to perform well to these specifications.

6.1.2 Rationale for Wafer-Bonding Technology

In the case of wafer stacking and thinning, standard wafer-level processing promises more efficient handling and thinning versus the single-die handling. Also, wafer-level processing allows for smaller TSVs and smaller associated TSV keep-out areas, which are needed for ensuring that stress around the TSVs will not affect device performance. This in turn results in significantly higher vertical interconnection density, thus achieving much higher bandwidth performance versus chip-level stacking. In addition, by designing the TSVs so that they directly connect to a wiring level on another wafer, the need of micropillars for the 3D interconnection can be eliminated in wafer stacking. This can lead to remarkable improvement compared to the limitations that are imposed by the pitch-scaling limitations of the rather large structures typically used for chip-level bonding. This provides another advantage when it comes to bandwidth potential comparison between these two technologies. It can be surmised that with respect to implementation in mainstream high-volume manufacturing, the characteristics of wafer-bonding technology point to two applications primarily. (1) The first is the massively parallel manufacturing of simple 3D cores and (2) the second is the 3D scaling of commodity memory, as in the case of memory cube type applications. In both cases, the prohibitively increasing costs, mainly due to lithography, and the yield and reliability issues associated with the scaling of 2D technology have made these two applications prime candidates for implementing this paradigm shift. It is also a great advantage that wafer-scale 3D stacking allows scaling by leveraging existing technology nodes by using chips of established technologies to achieve better performance, thus reducing costs, risks, and time-to-market potentially.

6.1.3 Description of Work

In this chapter, we describe two key technologies for chip stacking, (1) one achieved by use of die-level stacking and (2) another based on wafer-level stacking, with the main emphasis on the wafer-level stacking technology.

For the former, 3D die-level integration technology was developed using large-die 22 nm CMOS technology where die assembly was achieved on a laminate with seven layers of build-up circuitry on each core side. The design features integrated TSVs, microbumps, back-end-of-line (BEOL) wiring structures, and assembled controlled collapse chip connections (C4) joints, with Cu TSVs integrated in the BEOL. Void-free TSVs were formed, and the additional BEOL levels were fabricated after TSV processing and subsequently planarized by chemical–mechanical planarization (CMP). The types of bumps used to achieve interconnection were low solder volume with Cu pillars for the top die to the bottom die, and high solder volume for the bottom die to the laminate with the package exhibiting reliable assembly process, high-quality connections, and very good thermal performance.

For the latter, wafer-bonding process and integration technology was developed to achieve the stacking of high-performance POWER7™ cache cores [9] that were built based on 45 nm silicon-on-insulator (SOI) technology with embedded dynamic random access memory (EDRAM) and a total of 13 BEOL metal levels. For this work, copper TSVs of 5 μm in diameter were used at a 13 μm pitch for the signal communication and the power delivery to the stacked cache cores. The individual wafers, each featuring nine BEOL metal levels, were bonded permanently to each other by using a low-temperature oxide bonding. The topside wafers were thinned to about 13 μm using a combination of mechanical grinding, chemical–mechanical polish, and dry etching. Subsequently the TSVs were formed by using conventional lithographic alignment and definition techniques. Four additional metal levels were also built post-bonding and TSV definition to complete the interconnection of the chips in the two wafers and to enable testing. Electrical testing shows very good performance and device stability.

6.2 Chip-Level Three-Dimensional Integration with 22 nm Complementary Metal–Oxide–Semiconductor Technology

Realizing the advantages of the KGD benefits for chip stacking, a 3D chip-level integration technology was demonstrated that involves large Si die (>600 mm²) design, based on 22 nm CMOS technology with ultra low-k (ULK) BEOL at two different microbump pitch sizes, 61 and 131 μm [10,11]. A cross-section high-magnification optical microscope image of the package in Figure 6.1a shows the 3D integration with a 22 nm CMOS die stack using 61 μm pitch microbump on a laminate. Optical image of a bird-eye view of the 3D package is shown in Figure 6.1b. Die assembly was executed on a laminate with seven layers of build-up circuitry on each core side. This is a challenging integration scheme considering the effects of warpage in a TSV-featuring thin die, which makes the use of conventional furnace reflow problematic. With laminate substrates also undergoing additional warping during heating due to thermal coefficient of expansion (CTE) mismatch,

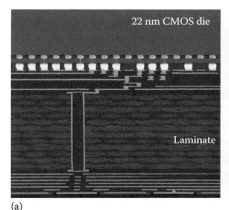
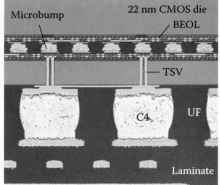

(a)

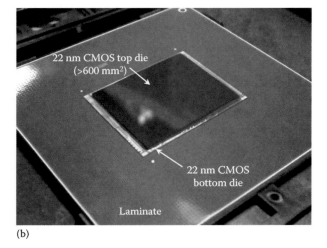

(b)

FIGURE 6.1
3D module with IBM's 22 nm CMOS die (>600 mm²) joined face to face with microbumps, TSVs, and C4s: (a) cross-section image and (b) optical image.

thin die warpage during 3D integration is increasingly challenging with an increase in the die size and a decrease in the respective component thickness [11,12].

These challenges, especially when it comes to maintaining thin die and laminate coplanarity, were addressed by using a new assembly technology [11]. More specifically, the Cu TSVs were introduced in the BEOL wafer manufacturing flow, with additional BEOL levels fabricated after TSV integration, before die singulation. Two types of bumps were used to achieve the desired interconnections during individual die stacking and packaging. The cross-section image in Figure 6.1a shows the integrated TSV, microbump, BEOL levels, and the assembled C4 joints to the laminate. As mentioned earlier, Cu-TSV is introduced in the BEOL by using a bottom-up

Cu electroplating process for TSV filling to ensure void-free TSV formation, without delamination of the BEOL dielectric stack. Additional BEOL levels were subsequently fabricated and planarized appropriately by using CMP.

Two types of bumps were used for the interconnection. Microbump (Cu pillar/SnAg solder) was used for the top die to bottom die bonding, whereas C4 was used for the bottom die to laminate bonding. The top solder system consisted of Cu and SnAg and was joined to Ni/Au pads on the bottom die side, whereas the bottom die side featured traditional SnAg bumps with terminal Ni joined to SAC 305 (Sn-3%Ag–0.5%Cu) presolder on the laminate side. The gap between the bottom die and the laminate is stable and no microbump/C4 joint failures were observed.

In order to assess the integrity of the interconnections at various locations, four-point measurements were used. More specifically, there were a total of 41 different locations of the module assembly in which four-point measurements were executed [11]. This included locations such as die centers but also potentially more challenging areas such as die corners. In Figure 6.2a there is a graph featuring the cumulative percent distribution of electrical resistance from these measurements. It should be noted that the resistance values obtained include contributions from the microbumps, the TSVs, the BEOL wiring, and the laminate wiring as well. There were two microbump geometries that were used for the connection of the bottom and top die. One was at a pitch of 61 μm and the other at a pitch of 131 μm, respectively. The main difference between the two cases was that the bump diameter was different and, as expected, with a different height for the pillars. The measurements suggest that the resistance of the smaller bumps at the smaller of the two pitches is about three to four times higher compared to the resistance for the larger bumps used at the longer pitch, which is not surprising based on the larger size of the bumps in the latter case. Based on the tight distribution exhibited by the measured resistance values, we can conclude that the assembly process is quite reliable, resulting in modules where the die-to-die interconnections are of good quality.

In order to validate the promise of this chip-level 3D integration technology, the large chip stacks with 131 μm pitch microbumps integrated with the laminate underwent thermal cycling (TC), temperature humidity bias (THB), and high-temperature storage (HTS) testing [10]. Table 6.1 summarizes the stress conditions that were used for the evaluation of the 3D modules and the corresponding test results. All stress times for these tests were performed according to the specifications of the JEDEC standard. The number of lots sampled and sample size per lot are also shown in Table 6.1. The test methodology included comprehensive continuity and leakage testing of a variety of macros that were specifically designed to assess the integrity of the upper-bump to TSV to lower bump connection, the laminate wiring to these structures, and the overall BEOL wiring level integrity in both chips in this package.

The thermal performance of the completed package is a key metric with regard to thermal reliability assessment. The interface from the top of the

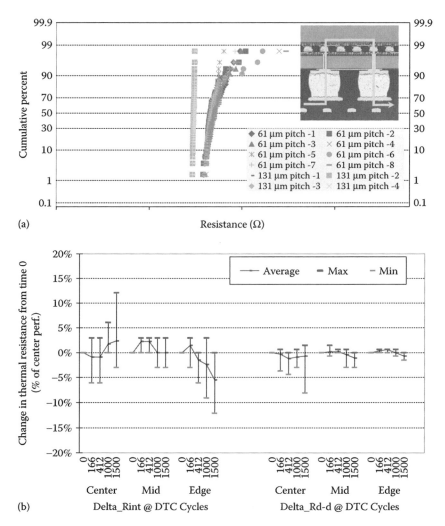

(a)

(b)

FIGURE 6.2
Reliability test results: (a) a cumulative distribution graph of electrical resistance including microbumps, TSV, BEOL wiring, and the laminate wiring, and (b) thermal reliability of TIM1 (R_{int}) and die-to-die (R_{d-d}) interfaces through 1500 cycles of DTC (−55/+125°C) accelerated stressing.

stacked die to the package lid through a thermal interface material level-1 (TIM1), also known as R_{int}, is monitored by implementing 25 top-chip sensors that are referenced to the external thermocouple attached to the lid. The sensors are classified based on the location on the die (center, middle, and edge) to track potential thermal performance degradation during deep thermal cycling (DTC) (−55°C to +125°C) by a TIM tearing, shearing, or loss of adhesion mechanism.

TABLE 6.1

Reliability Stress Conditions for Chip-Level 3D Modules and Test Results

Cell	Stress	Condition	JEDEC Spec	Requirements	Quantity (Lots)	Result
A	TC-K	0/125°C	A104	1000 cycles/0 Fail	31(4)	Pass
B	TC-G	−40/125°C	A104	850 cycles/0 Fail	13(1)	Pass
C	THB	85°C/85% RH/3.6V	A101	1000 hours/0 Fail	30(2)	Pass
D	HTS	150°C	A103	1000 hours/0 Fail	13(2)	Pass

A new thermal interface in the interconnection level between the die is unique to this stacked-chip structure. In this face-to-face interconnection configuration, this interface consists primarily of two BEOL stacks featuring bump and underfill material. Heat generated in the bottom die must transfer through this interconnection interface into the top die before it can travel through the R_{int} path to the exterior of the package. The key thermal resistance parameter characterizing this interface is resistance die-to-die (R_{d-d}). The R_{d-d} values are also classified as center, middle, and edge tracking groups, similar to the sensors.

Both R_{int} and R_{d-d} are calculated generally as follows:

$$R_{int}(C/W) = \frac{\left(T_i(\text{top die}) - T_{\text{thermocouple}}\right)}{\text{power}} \qquad (6.1)$$

$$R_{d-d}(C/W) = \frac{\left(T_i(\text{bottom die}) - T_i(\text{top die})\right)}{\text{power}} \qquad (6.2)$$

Thermal reliability is defined by the change in R_{d-d} or R_{d-d} over the duration of an accelerated stress test, such as DTC. In this case, a cumulative 1000 DTC was used with thermal readouts at the 166, 412, 1000, and 1500 cycle points. As shown in Figure 6.2b, both R_{int} and R_{d-d} exhibited extremely stable behavior throughout stressing. A maximum thermal degradation of <4°C per 1000 W of power dissipation was observed at the end-of-stress 1500 cycle readout.

6.3 Wafer-Bonding Technology for Three-Dimensional Integration Stacking

As described earlier, wafer-scale bonding can be a key enabler with regard to substantial benefits among which are interconnects (IC) with lower IC delay, much higher data bandwidth due to tighter pitches, and better power performance. From the manufacturing costs perspective, the much higher throughput is beneficial as multiple chip stacks can be formed in massively parallel fashion and then these stacks can be singulated during the subsequent process steps. This becomes possible with relatively few additions and modifications to established infrastructure in semiconductor fabs as long as the most suitable 3DI wafer-bonding integration scheme is utilized. We provide in what follows a brief review of main options for wafer bonding that are currently considered and we discuss the rationale for selecting oxide–oxide wafer bonding to achieve 3D integration of advanced CMOS for this work.

6.3.1 Metal–Metal Bonding

In the case of metal–metal wafer bonding, metal pillars, metal studs, or metal microbumps are created on the bonding surface side in each wafer first. Subsequently, proper alignment and wafer bonding establish the direct electrical interconnection between elements in the two wafers. There are many approaches for forming these metallic features that are needed for the metal-to-metal wafer bonding. Starting with the well-established method of using C4 interconnects, which is typically used in chip packaging [13], it quickly becomes obvious that it is not very suitable especially when it comes to attempting multiple wafer stacking. This is primarily not only due to the thermal budget constraints that it entails but also due to the large size of C4s themselves, which would of course lead to very large IC pitches and therefore would limit bandwidth severely [14]. One alternative method that is also being considered is using gold/tin (Au/Sn) soldering technology for the bonding. This involves microbump fabrication and is based on the nature of the eutectic compositions of Au/Sn metallurgical systems [15]. This can definitely provide an improvement versus C4s as it achieves direct electrical connection with low soldering temperature processes. It also possesses the added benefit of achieving self-aligning during the bonding process. Furthermore, it exhibits very good wetting behavior and reasonably adequate resistance against corrosion. Still, there can be concerns with the mechanical robustness and the stability of the bonded systems and for this reason, it typically requires underfill processes to enhance the robustness [16]. This of course increases the cost and complexity, and when combined with the fact that alignment performance, critical dimension (CD), and pitch scaling of interconnections can be challenging, it limits the possibilities for wide implementation for wafer bonding. As it is limited by the microbump dimensions and with throughput also being rather limited, it does not appear to be suitable for high-volume manufacturing.

Solid–liquid interdiffusion (SLID) is another metal-to-metal wafer bonding method where a higher melting point metal, such as Cu, is combined with a lower melting point metal, such as Sn. This method relies on the placement of the lower melting point metal in between the higher melting point metal stud structures [17,18]. A subsequent thermo-compression bonding process ensures the melting and diffusion of the lower melting point metal atoms into the higher melting point metal atoms, thus forming stable intermetallic states. However, this technique also is of rather limited scalability with respect to multiwafer stacking due to the inadequate thermal stability of the bonds compared to what is actually needed to handle the rigors of downstream wafer-level processing. The direct metal-to-metal thermal compression bonding is, on the other hand, more promising for wafer-scale bonding compared to other metal-to-metal wafer-bonding techniques. In this technique, the metallic structures, such as Cu microstuds and pillars, are built such that they protrude out of the wafer surface, typically after a recess process for the surrounding dielectric is done [19]. To prevent corrosion of these structures, it is essential to achieve clean metal surface before the thermo-compression bonding. For this purpose, typically clean processes are used to remove the surface oxides and any other impurities that may be detrimental to the bonding quality over time. By using this method, the advantage of direct electrical connection is retained, whereas high bond energy is achieved for the metal–metal bond interfaces with stability that can be enough to withstand downstream wafer-level processing. Having said that, underfill is typically needed to guard against mechanical stability and reliability risks, and this impacts the complexity and cost negatively. To ensure tight pitches and CD, the bonding alignment overlay performance on a wafer scale must be very accurate and more so in the case of smaller CD interconnect features. Another challenge is derived from the thermal and compression stress involved in the bonding process itself, which can affect alignment performance, process yield, and is also typically of relatively low throughput, which makes implementation for high-volume manufacturing less probable.

6.3.2 Hybrid Bonding

In the case of hybrid-bonding wafer-bonding processes, the bonding of the metallic interconnect pillar features on the wafer surfaces (as described in metal-to-metal bonding earlier) is also accompanied by the bonding of the surrounding dielectric surfaces, which can be polymer or more traditional dielectric materials [20,21]. If a dielectric is used, the most common case is the use of silicon oxide or silicon oxide-based interlayer dielectric (ILD). In the case of polymeric adhesive materials, a common material used is benzocyclobutene (BCB) adhesive. To achieve the simultaneous bonding of these surfaces, specific surface clean and surface preparation and planarization steps are combined, such as chemical mechanical polishing, plasma cleans, and wet cleans for surface cleaning and surface activation. An essential requirement for the successful hybrid bonding is achievement of low-wafer topography with atomically smooth roughness of the bonding surfaces. It is also quite common to use recessed metal structures with respect to the surrounding dielectric. In hybrid wafer bonding, there is typically an initial room temperature-bonding step to preserve bonding alignment overlay, which is followed by the main bonding process, the thermo-compression bonding step. This allows the recessed metal features to expand and bond with such opposing structures in the other wafer or in the case of polymer/metal hybrid bonding; it ensures bonding of both polymer–polymer and metal–metal surfaces.

The main advantage of this method is that direct electrical connection is achieved with a high bond strength across the surface of the wafer, thus allowing for robust bonding and therefore adequate resistance to downstream process stress, both mechanical and thermal. As an additional benefit, underfill is not required, which is a key differentiation versus metal–metal bonding techniques due to the increased mechanical stability achieved. The main challenges with hybrid bonding are primarily with respect to the bonding overlay and alignment performance, which must be very accurate and more so for increasingly smaller CD features [20,21]. A key concern is also with the reliability performance as Cu surfaces are not encased in impervious barriers during the simultaneous bonding of metal and insulator surfaces. Thermal and mechanical stresses inherent to the bonding process can also be a concern for process yield in the case of polymer/metal hybrid bonding. Overall, throughput is relatively low, especially if long thermal compression steps are involved to ensure strong bonding.

6.3.3 Oxide Bonding

Oxide wafer-to-wafer bonding requires the formation and preparation of silicon oxide-bonding layers on the wafers to be bonded to each other. These bonding layers are then typically activated and cleaned by using plasma and aqueous solution cleaning treatment. This is immediately followed by the wafers being aligned to each other, contact being initiated, and then releasing the wafers to spread this initial bonding and to preserve the alignment. This is subsequently followed by a brief post-bonding batch anneal process that ensures full-strength covalent bonding between the wafers at the oxide-bonding interface [7]. Wafer-scale integration using oxide-to-oxide bonding promises some very desirable properties that make it a very interesting choice as an upfront solution when it comes to high-volume manufacturing and with low-reliability risks [7,22].

To begin with, the use of oxide-bonding layers provides flexibility in terms of overcoming any topographic challenges on the surface of the wafers as their preparation and planarization can be tailored accordingly to accommodate incoming wafer topographies. The throughput of the oxide-bonding process compared to thermo-compression bonding can be almost an order of magnitude higher. Furthermore, the bonding alignment performance is generally quite accurate and stable as the initial bonding, which locks the alignment at room temperature. This technique also does not entail the significant thermal and mechanical stresses that are inherent in thermo-compression bonding. It should be noted that bumpless interconnect formation that is enabled provides not only a significantly higher IC density for signal transmission and higher power distribution but also more efficient thermal dissipation. One key disadvantage is that it does not provide for direct electrical connection, compared to the methods described earlier. But this can actually become an advantage, as this means that risks with respect to reliability (compared to the metal-to-metal or hybrid bonding) are minimal if any, as the interconnections are defined later through oxide layers. The one great advantage of oxide bonding is in the fact that it is much more robust when it comes to enabling standard wafer-level processing steps and with relatively high throughput. This is crucial when it comes to the pursuit of tighter IC scaling of CD and pitch through TSV scaling, which can be much smaller than the typical microbump-based schemes.

6.4 Oxide-Bonding Technology for Embedded Dynamic Random Access Memory Stacking

6.4.1 Oxide-Bonding Layer Preparation

The key requirements for the reliable oxide-bonding technology are driven by the necessity for good-quality oxide films and surfaces [7]. More specifically, the bonding films must be dense enough to allow for adequate bonding site density, and they must be atomically smooth to avoid nanoscale voids in the bonding interface, and they must neutralize any short- and medium-range wafer surface topography enough to avoid the formation voids during bonding. In addition, they must be very clean to avoid bonding defectivity and voids around particles at the bonding interface. Bonding surface flatness and the absence of short- and medium-range topography are crucial for defect-free bonding.

Considering these requirements, a two-film oxide-bonding film stack solution was developed. Both the films in this stack were deposited by using plasma-enhanced chemical vapor deposition (PECVD). This was first deposited based on a tetraethoxysilane (TEOS) process and was optimized in terms of its thickness and the subsequent planarization so as to overcome any potential incoming wafer topography, as needed. There was also a specially designed annealing step that enhanced overall surface cleanliness that enabled good adhesion properties with respect to the subsequently deposited second layer. A CMP step polishing step was also used to achieve sufficient flatness, followed by a cleaning step before proceeding with the deposition of the second film. The second bonding layer was then deposited based on silane chemistry to serve as the primary bonding layer interface. Also for this film, special thermal annealing to improve cleanliness and a polishing step to ensure low-surface roughness were used, according to the demands for defect-free oxide bonding. Representative results from stylus profilometry measurements obtained from challenging areas on the wafer in terms of topography, both before and after the planarizing film over a 2 mm by 5 mm area, suggest that topography is drastically reduced from ~250 nm range over the scanned area to ~5 nm, as it shown in Figure 6.3a [7]. Furthermore, atomic force microscopy (AFM) was used to characterize the topmost bonding film surface after preparation is completed. Such characterization shows

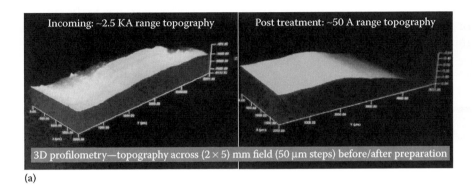

(a)

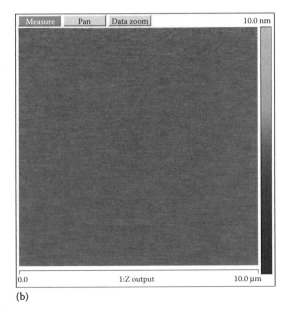

(b)

FIGURE 6.3
(a) 3D profilometry imaging results for a (2 × 5) mm field, in 10 nm steps, before and after the bonding film preparation over a challenging area of incoming wafer topography showing that topography is improved to ~5 nm over the scanned range. (b) Atomic force microscopy image showing the atomically smooth surface of the prepared bonding film stack.

atomically smooth surface with root mean square (RMS) roughness values between 0.2 and 0.4 nm, as demonstrated in Figure 6.3b. Thus, a smooth bonding surface is achieved as required for the bonding process. Prior work has shown the value of using the two-layer bonding stack system.

Although the first TEOS-based layer is thick enough to accommodate incoming topography, the bond strength that it can deliver upon bonding is not optimal. However, the use of silane-based low-temperature oxide as the topmost bonding layer provides better performance. The presence of defective regions in the silane-based oxide, where multiple Si-OH groups are concentrated, can accommodate H_2O that evolves later during the interfacial condensation reactions that take place during the bonding process and this results in higher bonding energy [23]. This interpretation is confirmed from the fact that prebaking of these films can increase their bonding energy, as any already absorbed H_2O is removed, leaving room for H_2O molecules generated later during the post-bonding anneal. The fact that the bonding energy is typically dependent on the N_2 plasma power for these films suggests that the surface density of Si-OH is key, and therefore silane-based films are more suitable than TEOS-based films.

6.4.2 Wafer-Level Three-Dimensional Integration with Oxide Bonding

The oxide bonding was executed by using a 300 mm wafer platform that was equipped with integrated wet clean module, plasma chamber, wafer pre-aligner, dual-microscope wafer aligner stage, anneal module, and vacuum thermal compression module. The bonding process involved the following steps: (1) loading the host and donor wafers, respectively, (2) activating the oxide-bonding layers by use of dual radio frequency (RF) nitrogen plasma, (3) cleaning the top oxide surface on both wafers via ultrasonication in aqueous solution, therefore terminating the activated hydrophilic surface with silicon to hydroxyl bonds, (4) loading the wafers to the aligner stage chucks with the two bonding surfaces facing each other, (5) aligning the host (bottom) wafer and the donor (top) wafers by detecting alignment key marks patterned on the wafers during the last metal-level fabrication through the calibrated dual aligner microscopes, (6) bringing the two wafers in close proximity with the bonding surface layers facing each other in the <10 μm range, (7) initiating contact from the center of the top wafer through a small piston pin assembly on the backside of the wafer, (8) releasing the donor wafer from the top chuck to conclude the initial bonding, and (9) waiting briefly for the van der Waals initial bonding to stabilize and preserve the alignment before returning the now bonded wafer pair into the wafer cassette for subsequent processing downstream.

The bonded wafer pairs were then batch annealed in a furnace at temperature ranges compatible with BEOL-processing specifications, typically <350°C, for 2 hours to ensure the full-strength covalent bonding at the bonding interface. Characterization of the bonding performance of this process has consistently shown void-free bonding as exhibited by scanning acoustic microscopy (SAM). Equally important is the fact that the bond strength, as characterized by the Maszara method [24], is typically in excess of >2 J/m², as it indicates that sufficient strength exists so as to withstand the rigors of downstream processing, especially the wafer thinning by grinding and polishing and any thermal deposition and anneal process-related thermal stresses. The typical alignment overlay performance of the bonding as determined by infra-red through-wafer microscopy measurements remains unchanged after the bonding anneal and is below 2 µm of the target throughout the bonded wafer chips, in many cases even below 1 µm. This is important as it enables the design of relatively small landing pads for the interwafer TSV connections, therefore keeping the real-estate requirements for these structures at a minimum, which is beneficial in terms of pitch and interconnect density [7] (Figure 6.4).

The specific purpose of this work is to showcase the suitability of oxide bonding as a flexible approach to build prototype-stacked devices and to evaluate key device and interconnect performance elements. Figure 6.5 features a schematic representation of the summarized wafer-level 3D integration process implemented, showing the use of a silicon carrier wafer to stack a thinned device wafer on a bottom full-thickness device wafer [4]. As a first step, a blank silicon wafer, intended to be used as handler for a device wafer, was bonded to a first device wafer by using the low-temperature oxide-bonding method described earlier. This enabled the subsequent thinning of the device wafer from its backside to a thickness of 10–13 µm by

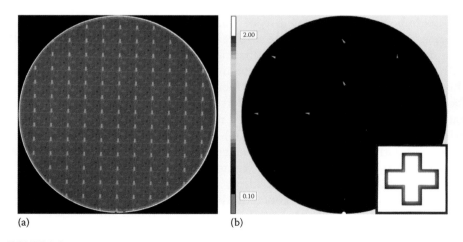

(a)   (b)

FIGURE 6.4
(a) Scanning acoustic microscopy image showing the absence of voids in the bonding between two wafers with an edge area of the bonded wafer system magnified to show virtually no defectivity. (b) Wafer-bonding overlay map based on measurements from cross-in-box structures from 13 chips showing bonding overlay performance <2 µm. The insert is an image of the overlay structures used.

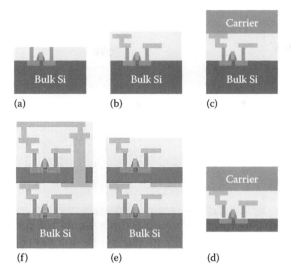

FIGURE 6.5
Schematic representation of wafer-level 3D integration process, showing the use of a silicon carrier wafer to stack a thinned device wafer on a bottom full-thickness device wafer. (a) FEOL, (b) BEOL, (c) carrier attach, (d) thinning, (e) bond, carrier removal, and (f) TSV and BEOL.

using a serial combination of mechanical grinding and polishing, reactive ion etching (RIE), and wet chemical thinning. All these processes were executed by using conventional silicon fabrication equipment and processes. Immediately after that, another oxide layer was prepared on the thinned backside of the device wafer and this wafer stack was bonded to another full-thickness device wafer, with the blanket handler wafer now at the top of the stack, therefore making sure that the thinned device wafer was bonded face up versus the full-thickness device wafer, with both device wafers in the same orientation, that is, face up. In the next step, the handler wafer was removed from the stack by a similar combination of mechanical grinding and polishing, RIE, and wet chemical thinning as the one described earlier.

Once this stacking and handler-removal processes were completed, copper TSV interconnects of a diameter of 5 μm were formed through standard lithographic definition, and a deep RIE etching process was executed from the front side of the thinned device wafer, which was designed to specifically address the ILD stacks, bond interface, and thinned silicon strata. After a post-RIE wet clean, the TSVs were lined with a conformal oxide insulator liner using a subatmospheric chemical vapor deposition (SACVD) process. Then another RIE step was utilized to ensure that the TSV insulator at the bottom of the TSVs would be removed to access the metal-landing pad underneath. The wafers then underwent the TSV metallization process, which includes metal liner and Cu seed deposition using physical vapor deposition (PVD) processes and a specially designed bottom–up copper-plating process

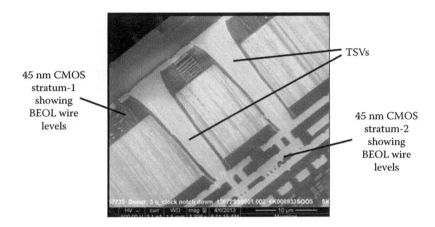

FIGURE 6.6
SEM cross section of donor and acceptor device wafers after wafer-level bonding and removal of handle wafer and subsequent TSV formation.

to fill the TSVs without voids. As a result, void-free TSVs were formed successfully with a 5 μm CD at a 13 μm pitch, as shown in Figure 6.6 [4], where a scanning electron microscopy (SEM) cross section of the joined and TSV-interconnected donor and acceptor device wafers is shown. This was followed by the fabrication of three additional copper-wiring levels on the top thinned device wafer. It is essential for high-performance, low-voltage applications to use low-resistivity Cu TSVs to limit the IR voltage drop. Throughout the whole process flow that enabled the wafer-level integration, all processing and tools utilized were compatible with high-volume manufacturing.

We should also note that we have demonstrated that this face-to-back wafer-level bonding integration process can be repeated to create interconnected wafer multistacks by adding more thinned wafers in the same fashion as the second wafer. With respect to the extendibility of this technology to include smaller diameter TSVs and thinner Si strata, we have successfully demonstrated four-wafer strata structures that feature 1 micron diameter interwafer TSVs chain structures at a Si strata thickness of ~6 μm, where each wafer also features intrawafer TSVs of 0.25 μm diameter, to connect the front and back side of the same wafer [8]. As can be seen in this prototype structure, SEM cross-section image shown in Figure 6.7a, a handle silicon wafer, four strata of wafers with both interwafer and intrawafer TSVs, and associated TSV chain structures were demonstrated. For each stratum, the intrawafer TSVs that eventually connect its front to its back are fabricated when the wafer is at full thickness from the front side. These intrawafer TSVs are of 0.25 μm in diameter. Once these are connected to the respective thin wire levels on the front side of the wafer, the bonding layer is prepared and then bonding ensues to the handler wafer (for the first stratum)

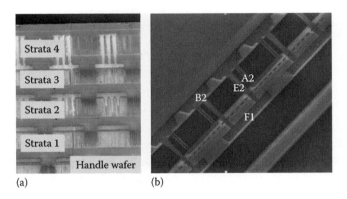

FIGURE 6.7
(a) SEM cross section of donor and acceptor device wafers after wafer-level bonding and removal of handle wafer and subsequent TSV formation. (b) SEM cross sections showing the formed TSV chain structures from wafer to wafer for evaluation for TSV performance.

or on stratum $n-1$ for the n stratum. In each case the newly bonded wafer is thinned to <7 μm to reveal the thin intrawafer TSVs from its back side. For $n > 1$ strata, interwafer TSVs of 1 μm in diameter are fabricated from the backside, using a process that etches the remaining silicon of the stratum and its complex front-end-of-line (FEOL)/BEOL dielectric stack on the front side, the bonding layer, and the ILD stack on the back side of the $n-1$ stratum. Subsequently, backside wiring is defined to complete test chains that are formed for the intra- and interstrata connections, respectively, as shown in the SEM cross-section image in Figure 6.7b. The low resistivity of Cu achieved through the special TSV-filling processes, especially the intrawafer TSVs, is extremely powerful and essential for the performance requirements with regard to the key potential application of multistacked DRAM memory. This TSV design system can allow for an unprecedented interconnection density, with very tight allowable pitches. This opens a very promising integration route toward high-volume manufacturing low-power, high-density 3D DRAM memory stacks.

6.5 Key Electrical Metrics Performance Results for Embedded Dynamic Random Access Memory Stacking

The wafer-stacking processes involve thermal processes up to 350°C; due to the differential thermal expansion mismatch present among metal/dielectric/silicon layers, this results in significant stress on the devices and structures, a serious concern for structural and parametric integrity. Furthermore, modern CMOS technologies involve strain-engineered elements that are sensitive

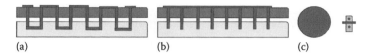

(a)   (b)   (c)

FIGURE 6.8
To accurately capture the effect of 3D stacking and TSV processing (a) TSV chains, (b) TSV banks, and (c) FETs in proximity to TSV were used.

to stress. So any significant perturbation resulting from 3D integration needs to be characterized thoroughly via structures that are specifically designed to capture the effect of these perturbations. Specifically, we utilize TSV chain structures that interconnect different strata (Figure 6.8a), TSV bank structures that join TSVs in parallel (Figure 6.8b), and structures that place field effect transistors (FETs) in proximity to TSV (Figure 6.8c). These structures, while necessary for capturing the impact of the bonding and TSV processing, are not sufficient to capture the full performance characteristics of the 3D stack that requires functional testing of memories at speed.

For this work, we focused on the electrical characterization of the two-wafer stack prototype featuring high-performance POWER7™ cache cores that were built based on 45 nm SOI technology with EDRAM [9]. As the resistance of one TSV is quite small, ranging in the milli-ohms, to accurately characterize the TSV resistance, chain structures comprising of multiple links, with each link containing two TSVs and interconnect wiring, were used. The median of the resistance for a specific chain structure was plotted against the number of links in the TSV chains as shown in Figure 6.9a. The linear behavior exhibited in the plot indicates that there is a consistent ohmic behavior, which is expected, with an average resistance of 120 mΩ per link. This value incorporates both the TSV and the local wire circuit resistance contributions. This resistance value is quite compatible with

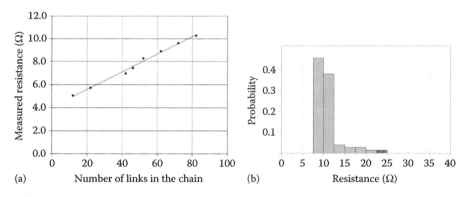

(a)   (b)

FIGURE 6.9
(a) Plot of measured median chain resistance versus number of links in TSV chain. The typical resistance per link was 120 mΩ, including the link wiring resistance. (b) Frequency plot of number of dies measured versus measured resistance for a TSV chain consisting of 82 links.

high-performance applications that typically require current density values in the order of 1 Amp/mm² or even higher [25,26]. The wiring impact from the necessary keep-out zones was estimated to be less than 1%. The distribution of resistance from a chain comprising of 82 links from several dies is shown in Figure 6.9b. The observed distribution is characterized by a standard deviation of roughly 10 mΩ per link, indicative of a well-controlled TSV-formation process.

In a 3D stack the TSV is utilized for power and signal transmission. The ability of the TSV to carry high-frequency signals is characterized by its impedance that is mainly capacitive. TSV banks comprising of different numbers of TSV's, joined in parallel, were constructed and their leakage and the capacitance characteristics were measured. Figure 6.10 shows the cumulative distribution of TSV banks with different numbers of TSV links. The capacitance scales with the number of links with an individual link capacitance measured is in the range of ~80 fF, a value which is significantly less than the case of bump-bonding technology [26,27]. The RC characteristics of these TSVs are consistent with the typical 100–200 μm length wiring load needs, which means that 3D macro-to-macro signaling can be achieved easily in this RC performance space and without having to resort to any buffering solutions. The leakage-to-substrate of the various TSV arrays was also measured and plotted for different numbers of TSVs in the respective capacitance banks as shown in Figure 6.11a and was also found to scale linearly with number. The observed leakage is very close to the lower limit of the tester capabilities, which is a very good sign regarding overall leakage performance.

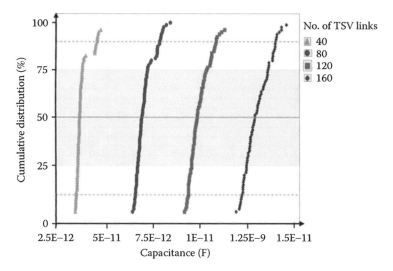

FIGURE 6.10
Plot of measured TSV bank capacitance versus number of TSVs in the bank. The capacitance measured per TSV was extracted as 80 fF per link.

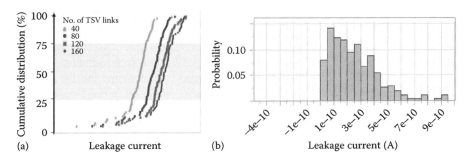

FIGURE 6.11

(a) Plot of measured leakage current versus number of TSVs in bank. The worst leakage per TSV was well below the pA level, suitable for most demanding CMOS applications. (b) Frequency plot of the distribution of the leakage current from a bank with 120 TSV.

The calculated current leakage per TSV is 1.18 pA at a 2V load, which means that for a 400 mm² chip design, which may feature more than 50 thousand TSVs, the total TSV leakage current would be approximately 59 nA. The distribution of leakage current measured, captured in Figure 6.11b, confirms consistent values in a very low range and practically negligible when it comes to most high-performance applications.

One key metric to verify that the devices are not perturbed by wafer bonding and the TSV integration process is to evaluate Idlin/Ioff characteristics for the FET devices in close proximity to TSV. In this respect, Ioff (Figure 6.12a) and Idlin (Figure 6.12b) were measured respectively for devices in proximity to TSV and devices far from TSV (control devices). The data were repeated before and after bonding. As it is apparent in Figure 6.12a and b, the FET devices Idlin/Ioff show no significant changes post wafer stacking and TSV formation and metallization versus the measurements on the single 2D wafers [22]. The TSV-induced Ion/Ioff variation

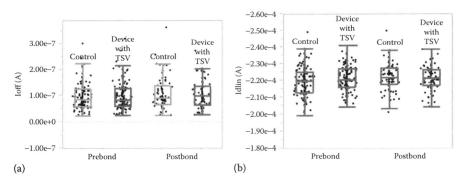

FIGURE 6.12

(a) Plot of Ioff for FET devices, control and devices with TSV in proximity. The data before bonding and after bonding showed no effect from either introduction of TSV or bonding. (b) Same as (a) but Idlin, a sensitive metric of TSV, process effects is shown.

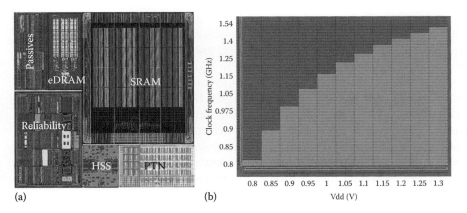

(a)   (b)

FIGURE 6.13
(a) The chip layout of a one strata. This stratum is joined to another identical stratum via 3D stacking (b), which shows the lowest functional frequency versus Vdd shmoo for eDRAM read and write operations demonstrating the functionality of a stacked eDRAM.

for NMOS transistors was not measured for the purpose of this work but it is expected to be smaller than P-type metal–oxide–semiconductor (PMOS) devices as demonstrated previously [27,28].

Through this wafer-level integration process, a high-performance 45 nm SOI stacked chip (layout is shown in Figure 6.13a) with stacked EDRAM cache prototype was built with more than 11,000 integrated TSVs. The chips in the two different strata are designed to emulate a stacked processor and cache chip assembly, as described in a previous work [14]. A built-in-self-test (BIST) engine was designed for the purpose of testing the strata-to-strata communication performance and to test the memory functionality of these structures for each stratum. The BIST engine successfully accessed the EDRAM on both strata of this prototype. The resulting shmoo plots, Figure 6.13b, of clock frequency versus supply voltage demonstrate 16 Mb EDRAM functionality, which is fixable, and points to possible strata-to-strata communication frequencies of up to 2.1 GHz at 1.3 V. The memory patterns for these tests were written in this stacked EDRAM prototype using four different configurations, as described in the following, and are also shown in Figure 6.12 [22]:

1. Single 2D-thick wafer mode where the memory on the thick wafer was activated.
2. Bonded 2D-thin wafer mode where the memory on thinned wafer S1 was activated and the test patterns were loaded using the TSVs.
3. 3D mode where the BIST engine on the thick wafer controls the memory on both wafer strata.
4. 3D mode where the BIST on the thin wafer controls the memory on both wafer strata.

By evaluating modes 3 and 4, the ability to write and read data from alternating strata memory in a single cycle was demonstrated, which confirms the quality of the clocks, power, control, and data signals across the chip boundary. This is necessary to be able to transfer the data at speed for the entire memory assembly without any errors. Furthermore, it is obvious from the shmoo plots in Figure 6.13 that the failure mechanism is shared by all the modes evaluated, all plots are practically identical, which indicates that the EDRAM performance was not affected by the wafer-level bonding and integration of the TSVs to achieve interconnection. The pattern shmoo that was used was the march 9 pattern, which by design forces cycle-to-cycle simultaneous switching patterns across the strata boundary for this evaluation. In addition, an equivalent column march pattern was also evaluated, with the results being very similar as well. The maximum allowed frequency at wafer test was limited by the voltage drop, which is inherent in the cantilever-probing method used, compared to socket-based module test. The retention signature obtained for the EDRAM indicates retention times achieved that were over 200 µs.

6.6 Summary

Two different technologies of 3D technology for chip stacking were demonstrated, both featuring very promising results. The first is a die-level stacking integration scheme featuring 22 nm CMOS technology with ULK BEOL, which enabled die assembly on a laminate with seven layers of build-up circuitry on each core side. These featured BEOL-integrated TSVs, microbumps, BEOL wiring, and C4s used for the bonding and interconnection were low-volume SnAg solder with Cu pillars for the top die to the bottom die and C4 for the bottom die to the laminate. Interconnect characterization showed high yield and high integrity of microbumps, TSVs, and C4 joints. Very good thermal performance was exhibited, with the test structures passing TC-K, TC-G, THB, and HTS, thermal stress test comfortably, which makes this a very good candidate for die-level stacking and packaging.

As the main focus of this work and aiming at high-volume manufacturing of high-performance memory, a wafer-level integration technology was developed through oxide bonding stacking of high-performance POWER7™ 45 nm SOI technology cache cores with EDRAM. The measured TSV capacitance and resistance are compatible with high-bandwidth chip–chip communication and the TSV leakage performance of 1 pA/TSV is very good. The FET Ion/Ioff shows no significant change post stacking and TSV processing, and functionality of 3D-stacked EDRAM cache cores has been confirmed by successful writing of memory patterns at up to 2.1 GHz at 1.3 V. This technology is highly compatible with existing high-performance logic technology and thus it is a very promising candidate for high-volume production of 3D-scaled logic and memory applications, especially since the technology elements for (a) further aggressive scaling of TSVs have been demonstrated and (b) $n > 2$ wafer stacking, up to four-wafer stacking have been demonstrated, promising higher performance and adequate extendibility.

Acknowledgments

The authors would like to extend their gratitude to all current and former colleagues at IBM who were involved in the work of 3DI technology development. Special thanks for their work and support to Jonathan Faltermeier, Wei Lin, Troy Graves-Abe, John Golz, Pooja Batra, Douglas LaTulipe, Alex Hubbard, Richard Johnson, Allan Upham, Toshiaki Kirihata, Jeffrey Zitz, Eric Perfecto, William Guthrie, Marcus Interrante, Richard Langlois, Koushik Ramachandran, Matthew Angyal, Vamsi Paruchuri, Thomas Gow, Mukesh Khare, Daniel Berger, John Knickerbocker, Subramanian Iyer, and T.C. Chen.

References

1. R.H. Dennard, F.H. Gaensslen, V.L. Rideout, E. Bassous, and A.R. LeBlanc, Design of ion-implanted MOSFET's with very small physical dimensions, *IEEE Journal of Solid-State Circuits*, 9: 256–268, 1974.
2. S.S. Iyer, T. Kirihata, M.R. Wordeman, J. Barth, R.H. Hannon, and R. Malik, Process-design considerations for three dimensional memory integration. In *Proceedings of the Symposium on VLSI Technology*, Honolulu, HI, June 16–18, 2009, pp. 60–63.
3. W. Arden, M. Brillouët, P. Cogez, M. Graef, B. Huizing, and R. Mahnkopf, More than Moore white paper. In *International Roadmap Committee for the International Technology Roadmap for Semiconductors*, 2010. www.itrs2.net.
4. M. Koyanagi, H. Kurino, K.W. Lee, K. Sakuma, N. Miyakawa, and H. Itani, Future system-on-silicon LSI chips, *IEEE Micro*, 18 (4): 17–22, 1998.
5. K. Sakuma, P.S. Andry, C.K. Tsang et al., 3D chip-stacking technology with through-silicon vias and low volume lead-free interconnections, *IBM Journal of Research & Development*, 52 (6): 611–622, 2008.
6. K. Sakuma, P.S. Andry, C.K. Tsang et al., Characterization of stacked die using die-to-wafer integration for high yield and throughput. In *Proceedings of the IEEE Electronic Components and Technology Conference (ECTC)*, Lake Buena Vista, FL, 2008, pp. 18–23.
7. S. Skordas, D.C.L. Tulipe, K. Winstel et al. Wafer-scale oxide fusion bonding and wafer thinning development for 3D systems integration. In *Proceedings of the 3rd IEEE International Workshop on Low Temperature Bonding for 3D integration (LTB-3D)*, Tokyo, Japan, May 22–23, 2012, pp. 203–208.

8. W. Lin, J. Faltermeier, S. Skordas et al., Prototype of multi-stacked memory wafers using low-temperature oxide bonding and ultra-fine dimension copper through-silicon via interconnects. *Proceedings of SOI-3D-Subthreshold Microelectronics Technology Unified Conference (S3S)*, San Francisco, CA, 2014, pp. 152–154.

9. J. Barth, W.R. Reohr, P. Parries et al., A 500 MHz random cycle, 1.5 ns latency, SOI embedded DRAM macro featuring a three-transistor micro sense amplifier, *IEEE Journal of Solid-State Circuits*, 43: 86–95, 2008.

10. K. Sakuma, S. Skordas, J. Zitz et al., Bonding technologies for chip level and wafer level 3D integration. In *Proceedings of the IEEE Electronic Components and Technology Conference (ECTC)*, Lake Buena Vista, FL, 2014, pp. 647–654.

11. K. Sakuma, K. Tunga, B. Webb, K. Ramachandran, M. Interrante, H. Liu, M. Angyal, D. Berger, J. Knickerbocker, and S. Iyer, An enhanced thermocompression bonding process to address warpage in 3D integration of large die on organic substrates. In *Proceedings of the IEEE Electronic Components and Technology Conference (ECTC)*, San Diego, CA, 2015, pp. 318–324.

12. K. Sakuma, K. Sueoka, Y. Orii et al., IMC bonding for 3D interconnection. In *Proceedings of the IEEE Electronic Components and Technology Conference (ECTC)*, Las Vegas, NV, 2010, pp. 864–871.

13. K. DeHaven and J. Dietz, Controlled collapse chip connection (C4)-an enabling technology. In *Proceedings of the IEEE 44th Electronic Components and Technology Conference (ECTC)*, 1994, pp. 1–6.

14. P.A. Totta, S. Khadpe, N.G. Koopman, and M.J. Sheaffer, Chip-to-package interconnections. In *Microelectronics Packaging Handbook, Part II*, 2nd ed., Tummala, R.R., Rymaszewski, E.J., and Klopfenstein, A.G. (Eds.). Springer: New York, 1997, pp. 129–283.

15. H. Oppermann and M. Hutter, Au/Sn solder. In *Handbook of Wafer Bonding*, Ramm, P., Lu, J.J.-Q., and Taklo, M.M.V. (Eds.). Wiley-VCH Verlag GmbH & Co.: Weinheim, Germany, 2012, p. 119.

16. K. Sakuma, S. Kohara, K. Sueoka, Y. Orii, M. Kawakami, K. Asai, Y. Hirayama, and J.U. Knickerbocker, Development of vacuum underfil technology for 3-D chip stack, *Journal of Micromechanics and Microengineering*, 21: 035024, 2011.

17. L. Bernstein, Semiconductor joining by solid-liquid-interdiffusion (SLID) process, *Journal of the Electrochemical Society*, 113: 1282–1288, 1966.

18. N. Hoivik and K. Aasmundtveit, Wafer level solid-liquid interdiffusion bonding. In *Handbook of Wafer Bonding*. Wiley-VCH Verlag & Co.: Weinheim, Germany, 2012, p. 181.

19. K.N. Chen and C.S. Tan, Thermocompression Cu-Cu bonding of blanket and patterned wafers. In *Handbook of Wafer Bonding*. Wiley-VCH Verlag & Co.: Weinheim, Germany, 2012, p. 161.

20. L. Di Cioccio, Cu/SiO2 hybrid bodning. In *Handbook of Wafer Bonding*. Wiley-VCH Verlag & Co.: Weinheim, Germany, 2012, p. 237.

21. M. Nimura, J. Mizuno, K. Sakuma, and S. Shoji, Solder/adhesive bonding using simple planarization technique for 3D integration. In *Proceeding of the IEEE Electronic Components and Technology Conference (ECTC)*, Lake Buena Vista, FL, 2011, pp. 1147–1152.

22. P. Batra, D. LaTulipe, S. Skordas et al., Three-dimensional wafer stacking using Cu TSV integrated with 45 nm high performance SOI-CMOS embedded DRAM technology, S3S, 2013.

23. W. Lin, L. Shi, Y. Yao, A. Madan, T. Pinto, N. Zavolas, R. Murphy, S, Skordas, and S.S. Iyer, Low-temperature oxide wafer bonding for 3-D integration: Chemistry of bulk oxide matters, *IEEE Transactions on Semiconductor Manufacturing*, 27 (3): 426–430, 2014.
24. W.P. Maszara, G. Goetz, A. Caviglia, and J.B. McKitterick, Bonding of silicon wafers for silicon-on-insulator, *Journal of Applied Physics*, 64 (10): 4943, 1988.
25. M.G. Farooq, T.L. Graves-Abe, W.F. Landers et al., 3D copper TSV integration, testing and reliability. In *Proceedings of the IEEE International Electron Devices Meeting (IEDM)*, Washington, DC, December 5–7, 2011.
26. J. Golz, J. Safran, B. He et al., 3D stackable 32 nm High-K/Metal Gate SOI embedded DRAM prototype. In *Proceedings of the Symposium on VLSI Circuits*, Honolulu, HI, June 15–17, 2011, pp. 228–229.
27. A. Mercha, G. Van der Plas, V. Moroz et al., Comprehensive analysis of the impact of single and arrays of through silicon vias induced stress on high-K/metal gate CMOS performance. In *Proceedings of the IEEE International Electron Devices Meeting (IEDM)*, San Francisco, CA, December 6–8, 2010, pp. 2.2.1–2.2.4.
28. L. Yu, W.-Y. Chang, K. Zuo, J. Wang, D. Yu, and D. Boning, Methodology for analysis of TSV stress induced transistor variation and circuit performance. In *Proceedings of the 13th International Symposium on Quality Electronic Design (ISQED)*, Santa Clara, CA, March 19–21, 2012, pp. 216–222.

7

Toward Three-Dimensional High Density

S. Cheramy, A. Jouve, C. Fenouillet-Beranger, P. Vivet, and L. Di Cioccio

CONTENTS

7.1 Introduction

3D (three-dimensional technology) high-density integrations have gained an increasing interest in the objective of maintaining high performances and/or low-power consumption as the Moore's law was slowing down.

If high-end advanced packaging solutions are developed, mainly for heterogeneous packaging, those alternatives are not suitable to fulfill the requirements for power-efficient applications, such as CIS (CMOS [complementary metal–oxide–semiconductor] image sensor), high-performance computing (HPC), gaming, and data center.

As an example, back-side illuminated (BSI) imagers players have released since 2012 many 3D prototypes and some are already in production, mainly driven by mobile phone applications. 3D technology used for 3D BSI is based on direct hybrid bonding, a pitch below 10 µm is reachable, pitch that is not achievable with conventional advanced packaging techniques. The idea is to dedicate top and bottom wafer to image sensor and logic function, respectively, rather than integrating both functions on the same floor plan.

We can predict that this specific application will pave the way for other products; the 3D approach that will be used will depend on the granularity scale of the product partitioning, as shown in Figure 7.1.

Below 10 µm pitch of chip-to-chip interconnection, two complementary 3D solutions are taking over

1. A back-end-of-line (BEOL)-type technology, based on Cu/SiO$_2$ direct hybrid bonding in the range of a few micrometers of pitch.
2. A front-end-of-line (FEOL)-type technology, named CoolCube™ at CEA-Leti, in the range of few tens of nanometers of pitch.

The basis of these principles, process integration, challenges, and perspectives are described in this chapter. Besides, 3D parallel and 3D sequential both are coming with new and common challenges, among which thermal dissipation. Last paragraph of this chapter dedicated to 3D high-density integration will focus on thermal simulation.

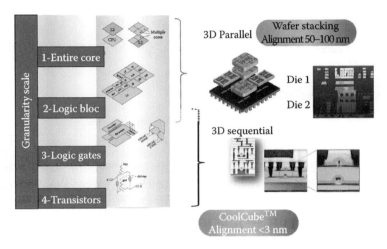

FIGURE 7.1
3D integration pitch roadmap, depending on design granularity.

7.2 Cu/SiO$_2$ Hybrid Bonding

This paragraph will detail processes and challenges of the BEOL-type high-density technology, based on the Cu/SiO$_2$ direct bonding.

7.2.1 Cu/SiO$_2$ Hybrid Bonding Principle

The Cu/SiO$_2$ hybrid bonding process is one variety of direct bonding technique, which refers to a process by which two mirror-polished wafers are put into contact and held together at room temperature by adhesive forces, without any additional materials [1]. For hybrid bonding, the interfacial materials are heterogeneous to create the electrical contact right after assembly and successive annealing.

This technology has become more and more attractive for microelectronic applications for its advantages, such as throughput and possibility of fine pitch of interconnection, compared to conventional flip-chip technologies. It can fit with 3D integration using wafer-to-wafer (WtW) bonding and die-to-wafer (DtW) bonding, which makes this process suitable for a large variety of applications.

Several metallic materials have been evaluated for hybrid bonding such as Au, W [2,3], but among the evaluated materials, Cu is the best candidate as it is already widely used in semiconductor device industry.

Direct bonding of patterned oxide/copper surfaces at room temperature and ambient air was developed by CEA-Leti since several years ago. The following paragraphs describe the characteristics of this integration illustrated thanks to most relevant results published in this field.

7.2.2 Technical Challenges Linked to Hybrid Bonding

This paragraph focuses on the main challenges to overcome to obtain a perfect direct bonding.

7.2.2.1 Surface Preparation

Wafer surface preparation before bonding is the key step for a successful direct bonding, whatever this bonding is patterned or not and whatever the 3D integration scheme (WtW or DtW). Therefore, to obtain defect-free hybrid bonding, the best choice is to use a standard BEOL process including chemical–mechanical polishing (CMP) step and then target a surface as flat as possible. Main challenges are elimination of typical defects such as dishing or recess of the copper pads on patterned surfaces, or array erosion as illustrated on Figure 7.2.

At CEA-Leti, an optimized CMP-based process was implemented on patterned surfaces to reach a high-surface planarization level [4]. The CMP optimization (pad/slurry matching) aims to minimize copper dishing and oxide erosion with respect to the layout. Number of CMP steps can be tuned to improve Cu/SiO_2 selectivity and reduce copper dishing: A three-step CMP process demonstrated excellent WtW-bonding performances with a CMOS imager application layout [5] but complementary work has shown that adding a step of copper protrusion could even improve topology management to enlarge process window and relax design rules for bonding levels [4].

Silicon oxide and copper roughness have been found to be lower than 5Å (Figure 7.3), which is in specification for excellent direct bonding [4–6].

After surface preparation, topology of pads with dimensions ranging from 3 to 6 µm depending on layout [7,8] has been characterized by atomic force microscopy (AFM) and shows a recess value that does not depend on pad dimension or shape (square or octagon).

It has been published that the CMP process could be completed by a plasma-prebonding treatment (N_2, Ar, H_2...) to increase adhesion energy

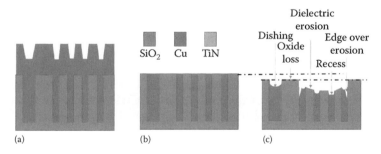

FIGURE 7.2
Principle of the damascene CMP. (a) Wafer post plating, (b) wafer post-CMP idealistic case, and (c) wafer post-CMP realistic case.

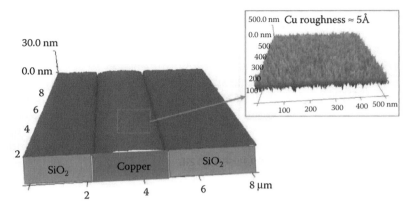

FIGURE 7.3
Post-CMP 3D AFM images showing flat oxide/copper interface and below 5Å roughness copper surface.

if required [9,10]. However at CEA-Leti, this treatment was not found to be mandatory and bonding was achieved right after CMP process.

7.2.2.2 Bonding Interface Characterization

After CMP process, both substrates are bonded together at room temperature, without applied pressure. It is then highly important to check the bonding interface quality to ensure following stages feasibility (such as backside steps for CMOS image sensors). Bonding quality is assessed through the following parameters:

- *Interfacial voids*: No interfacial voids should be detected.
- *Bonding strength*: Adhesion energy should be sufficient to support the steps following processing.
- *Top and bottom electrical pads alignment*: Mandatory to respect electrical specifications.

Interfacial voids in bonding interface can lead to electrical connection issues and then yield loss, as well as wafer breakage. Interfacial voids can be checked thanks to infra-red light or acoustic microscopy. This second characterization technology is widely used at CEA-Leti and in production as it enables a fast-complete scan of the wafer. The resolution of the void detection depends on the frequency of the scanning head, and with a 110 MHz transducer, the effective lateral resolution is 7 µm and defects as small as 5 nm can be detected. With this characterization mode, the variation in bonding quality is revealed by color contrast.

Nonbonded zones named *voids* may be observed at the interface:

- *Voids due to particles*: They are due to insufficient cleaning before bonding; they are in the range of few millimeters to several centimeters depending on particle size. For example, it has been demonstrated that a 1 μm particle can lead to a 5 mm width bonding void for a 525 μm thick substrate [11]. In the Figure 7.4, two acoustic scans coming from 300 mm bonding, presenting copper pads in the range of 3–6 μm, are shown. The bonding on the left presents some bonding defects, whereas bonding on right is perfect. The particle defect is the bright circle on the wafer.
- *Voids due to copper dishing*: We mentioned in our previous paragraph that non optimized CMP process may generate copper and/or oxide recess in the range of few nanometers (Figure 7.2). Yet this recess could lead to remaining voids at bonding interface and poor electrical connection as shown by b-type defect on Figure 7.4.

However, it has been demonstrated that bonding process could be tolerant to low topologies remaining after CMP process. This tolerance is obtained, thanks to copper expansion with temperature that enables the copper-to-copper connection (Figure 7.5).

Thus, the annealing step that occurs after bonding to strength the SiO_2/SiO_2 bonding is also used to compensate copper recess. Evolution of bonded structures with temperature can also be observed, thanks to scan acoustic microscopy (SAM). 200 mm-patterned wafers have been processed by previously described CMP leading to copper dishing and were subsequently bonded on blanket oxide wafers. These pairs were then successively annealed at 200°C and 400°C for 2 hours.

After the wafer bonding and successive annealing, the bonding interface quality is observed using a SAM. In Figure 7.6, the evolution with

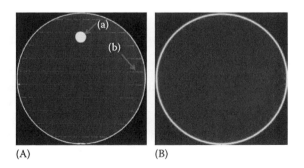

(A)   (B)

FIGURE 7.4
(A) Acoustic microscopic images of 300 mm Cu/SiO_2 bonding with defects: (a) large circular defect due to a particle and (b) smaller defects due to excessive copper recess. (B) 300 mm Cu/SiO_2 bonding without interfacial defects before annealing.

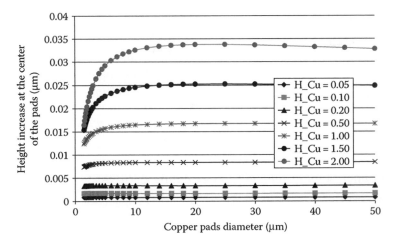

FIGURE 7.5
Ansys simulation of the vertical displacement of copper pads at 200°C, as a function of pad geometry. H is the height of the copper line in micron.

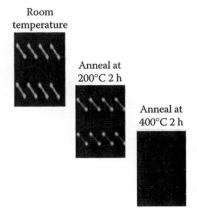

FIGURE 7.6
SAM images showing the evolution of half Kelvin cross patterns with temperature, the width of the branch is respectively from top to bottom 25, 25, 20, and 15 µm.

temperature, between room temperature to 400°C, of a SAM signal on a set of crosses of sizes ranging from 15 to 25 µm can be seen. At room temperature, the residual CMP dishing makes the copper areas unclosed: The whole crosses appear bright white. When temperature is increasing, the copper structures expand, filling the gap left by the CMP process. When the copper surface comes into contact with the facing oxide surface, bonding occurs and SAM contrast disappears (anneal at 400°C for 2 hours).

This experiment confirms the vertical copper displacement during post-bond anneal process and assess SAM relevancy for interfacial bonding qualification.

7.2.2.3 Bonding Energy Evolution with Temperature

Bonding strength can be determined using the double cantilever beam (DCB) technique [12]. This technique consists of measuring the debonding length induced by the insertion of a blade at the bonding interface. Figure 7.7 presents SiO$_2$/Cu wafers (copper density 20%) bonding toughness evolution depending on temperature in the case of 200 mm wafers. The bonding energy ranges from 1 J/m^2 at 200°C to 6.6 J/m^2 at 400°C, which is sufficient to successfully achieve the following stages of process.

7.2.2.4 Copper Modification with Temperature

Electrical performances will directly depend on the copper pads connection quality. Therefore, many studies have been carried out to investigate the copper-to-copper direct bonding mechanism [13–15], from the bonding interface at room temperature to its evolution with temperature. Therefore, copper-to-copper connection was firstly analyzed after bonding and annealing by focussed ion beam-scanning electron microscopy (FIB-SEM). An example of a cross-section of a 3.6 µm copper pad after bonding and annealing is shown in Figure 7.8.

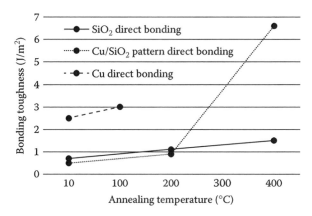

FIGURE 7.7
Comparison of bonding toughness for various wafer configurations. (From Di Cioccio, L. et al., An overview of patterned metal/dielectric surface bonding: Mechanism, alignment and characterization, *JECS*, 81–86, 2011.)

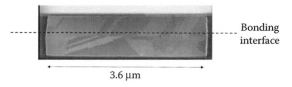

FIGURE 7.8
FIB/SEM cross-section of bonded Cu pads.

Neither large void nor copper recess can be detected.

For better understanding of copper-to-copper bonding mechanisms, complementary characterizations such as X-ray diffraction (XRD), X-ray reflectivity (XRR), and transmission electron microscopy (TEM) characterizations have been realized on wafer fully deposited with copper prior to bonding as described by P. Gueguen et al. Fully 200 mm copper-plated wafers have been treated with the previously described CMP recipe to ensure a low roughness and high hydrophilic behavior. TEM observations in Figure 7.9 have shown a sharp interface at room temperature into which a 4 nm crystalline layer could be seen at high magnification. Electron Energy-Loss Spectroscopy (EELS) analyses have detected the presence of oxygen in this layer which is thought to be the signature of the presence of a copper oxide at the interface.

At 200°C, the copper oxide becomes thermodynamically unstable. Its diffusion along the bonding interface is observed and diffusion bonding can happen.

Nanovoids partially filled with oxide are created in parallel to a copper-to-copper grain boundary growth, which is even more significant after 400°C post-bond anneal, as shown in Figure 7.10.

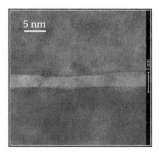

FIGURE 7.9
High-resolution TEM cross section of direct copper bonding without post-bonding anneal. A 4 nm thick crystalline layer is present at the bonding interface.

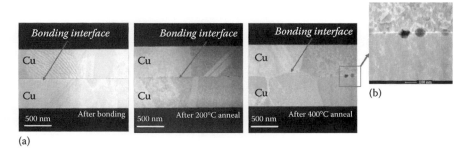

FIGURE 7.10
(a) STEM cross sections of direct bonding with successive post-bond annealing. (b) TEM observation of a typical interfacial void.

More recently, L. Di Cioccio et al. published a complete study to describe into details the void formation mechanisms in the case of three different metallic bonding: copper, tungsten, and titanium [15].

7.2.2.5 Wafer Alignment Consideration

As mentioned in introduction, Cu/SiO_2 hybrid bonding is shown to address interconnection pitch close to 1 µm; therefore, to recover vertical interconnect in 3D technology it is mandatory to bond with an alignment accuracy coherent with the desired interconnection dimension. Indeed, misalignment between top and bottom copper pads would lead to contact resistance increase.

To reach this goal, the WtW bonding equipment suppliers have been working for several years to develop manufacturing tools that match the interconnection roadmap. It is now demonstrated that alignment with 200 nm (3σ) over a full 300 mm wafer can be reached with a perfect control of translation, rotation, and scaling as shown in Figure 7.11 [6,7].

7.2.2.6 Investigation on Copper Diffusion

It is important to notice that a pad misalignment leads to areas where copper is directly in contact with oxide material without any diffusion barrier material, therefore one could fear a copper diffusion in the dielectric after the post-anneal bake that can go up to 400°C/2 hours. However, work achieved by Y. Beilliard et al. [16] has confirmed, thanks to energy-dispersive X-ray spectroscopy-analytical scanning transmission electron microscopy (EDX–STEM), that copper diffusion remains under detection limit, even after a thermal storage of the bonding at 300°C for 336 hours (Figure 7.12).

The Cu/SiO_2 interface stability has also been observed by S. Lhosthis et al. [8] after a 400°C 2 hours annealing.

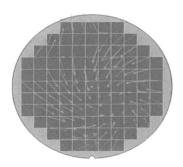

FIGURE 7.11
Vector map of overlay data quantified from a bonded 300 mm wafer pair using the EVG®40 NT measurement system. Max overlay here is 250 nm.

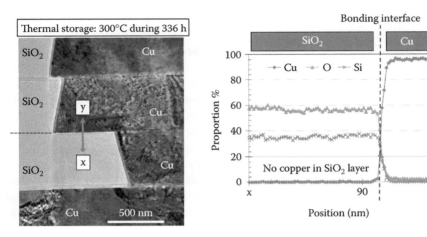

FIGURE 7.12
STEM cross-section image and EDX–STEM mapping of the misalignment area of hybrid bonding interface. The graph presents the Cu, Si, and O concentration profiles across the bonding interface.

However, it is important to note that specific pads can be designed to reduce alignment criterion. For example, Wang et al. recently communicated on a specific 25 µm bonding pad designed as a grid, which allows a 10 µm misalignment tolerance [9]. This is a very interesting strategy as long as the device connection density can be relaxed to this innovative larger pad size.

7.2.2.7 Pad Dimension Reduction

For industrial applications, density of interconnection will have to be continuously increased. Some recent publications have investigated feasibility of hybrid bonding at very fine pitch below 2 µm for simple integration (1 level of copper per wafer). Two recent publications have confirmed that even if the pad dimension is getting as large as conventional copper grain size (1.1 measured on 4.4 µm large pads) [8], hybrid bonding surface preparation processes and tool performances enable ultrafine pitch hybrid bonding [9,17].

7.2.3 Electrical Performances Evaluation of Hybrid Bonding

7.2.3.1 Electrical Structures Presentation and Performances

Over the past years, several publications from CEA-Leti and ST Microelectronics assessed the electrical behavior of different test structures by measuring their resistance under direct DC current [8,16,18–21]. Along with time and process maturity growing, the complexity of the evaluated structures has increased to be better representative of future industrial products.

The cross sections after CMP, bonding, and annealing of these structures are presented in Figure 7.13. The most recently published ones (case 3) consist of copper pads and via achieved thanks to dual damascene process.

Before electrical tests, all the bonding pairs have been characterized with the previously described protocol: No void has been detected, as shown on the perfectly bonded 300 mm wafer presented in Figure 7.13. This confirms the robustness of CMP-based surface preparation process.

In order to enable the electrical tests, a backside electrical contact has to be created on top wafer. Therefore, the bonded pairs have all been thinned successfully at various thicknesses: 50 μm with through-silicon via (TSV) and under bump metallization (UBM) in case 1; 40 μm in case 2 with further Ti/TiN/AlSi redistribution layer (RDL) and electrical pads, and more recently down to only 3 μm on a 300 mm wafer with following aluminum pads creation process (case 3).

Table 7.1 presents a comparison of the specific contact pad resistances measured on the three integration schemes presented in Figure 7.13.

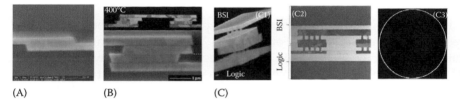

(A)   (B)   (C)

FIGURE 7.13
Contact pad cross-sections of the three different hybrid bonding stacks evaluated by CEA-Leti and STMicroelectronics. Case 1 and 2 are 200 mm wafers, Case 3 is 300 mm wafers. (A) Case 1: 1+1 copper level (From Taïbi, R. et al., *ECTC proceedings*, 219–225, 2010; Taïbi, R. et al., *IEEE IEDM*, 2011), (B) Case 2: 2+2 copper level (From Beilliard, Y. et al., *IEEE 3DIC*, 2014) and (C) Case 3: 3+3 copper level. (Lhostis, S. et al., *IEEE 66th ECTC*, Las Vegas, NV, 2016; Moreau, S. et al., *IEEE 66th ECTC*, Las Vegas, NV, 2016; Jourdon, J. et al., *IEEE IRPS*, 2017.)

TABLE 7.1

Specific Contact Resistance Comparison of Electrical Structures with Different Numbers of Copper Layers

	Case 1	Case 2	Case 3
Number of copper levels (top + bottom)	1 + 1	2 + 2	3 + 3
Seed layer material	TiN	TiN	TaN/Ta
Smallest pad size	3 × 3 μm	3 × 3 μm	3.6 × 3.6 μm
Wafer diameter	200 mm	200 mm	300 mm
Specific contact resistance @400°C	**22.55 mΩ·μm²**	**154 mΩ·μm²**	**130 mΩ·μm²**
Longest DC chain	29,422	30,160	30,000

The specific contact resistance ρ_c is representative of the resistance induced by the hybrid bonding interface and can be calculated with the following formula:

$$\rho_c = R_c \times A_c$$

where:
   R_c is the contact resistance
   A_c is the contact area

This resistance is extracted experimentally from the difference between theoretical calculations and experimental measurements of the daisy chain resistances.

It can be seen from Table 7.1 that the contact resistance induced by direct copper-to-copper bonding is very low. For example, R. Taïbi et al. have demonstrated that the contact resistance of 3×3 cm^2 pad can be as low as 2.5 mΩ [18], which confirms that a copper-to-copper connection obtained with hybrid bonding is less resistive than more classical µbumps (copper + solder), which presents a total resistance of more than 50 mΩ (without considering the connecting copper lines) [22].

Process maturity increase has recently confirmed that connection yield could be excellent with hybrid bonding on 300 mm wafers [8]. For example, the mean unitary resistance of a 30,000 connection daisy chain with 3.3×3.3 µm^2 square pads is only 82 mΩ (including connection copper lines) with a standard deviation of 6 mΩ over a set of more than 100 tested wafers. Connection yield is close to 100%. This value is lower than the unitary resistance of a daisy chain with 10 µm width µbumps connection, which has been found to be over 100 mΩ [22].

7.2.3.2 Environmental Reliability Study

In order to investigate the environmental reliability of hybrid bonding interface, all the bonded wafers have been exposed to a variety of reliability tests including environmental reliability as unbiased Highly Accelerated Stress Test (HAST) (uHAST, JESD22-A118A), Temperature Cycling (TC, JESD22-A104D), Temperature Storage (TS, JESD22-A103C), and ElectroMigration (EM) tests.

Unbiased HAST tests have been achieved on a 416-connection daisy chain packaged and wire-bonded after dicing [16] or at wafer level on backside aluminum pads [8]. In both cases, no significant variation of daisy-chain resistance has been observed after unbiased HAST test (85°C/85% RH/168h), TC (−50°C/+150°C), and thermal storage (1000 hours +1000 hours).

High-temperature thermal storage or thermal cycling resistance have been evaluated at wafer level on 300 mm diameter wafers [8]. In this case the performance of the 30,000-connection daisy chain with 3.6 µm pads (7.2 µm pitch) has been followed. A very small decrease in the electrical resistance of the daisy chains is observed after High-Temperature Storage (HTS). The maximum difference is −2% after 2000 hours storage, far from the −10% difference of the reliability specification. This resistance decrease is attributed to a better copper rearrangement in the lines and via or at the hybrid bonding interface with temperature. No significant evolution of the leakage current and capacitance of the combs is observed after thermal storage.

7.2.3.3 Electromigration

EM tests were performed on three structure types to identify possible failure related to hybrid bonding. This paragraph will focus on results with more complex structure including copper via and pads (case 3 Figure 7.13). In this case, EM has been studied using NIST (National Institute of Standards and Technology)-like or 100-connections daisy chain (DC100).

For NIST-like test structure, failure analysis reveals voids that are always in the single damascene line of the top/bottom wafer depending on the electron flow direction (up/downstream).

For DC100, no EM-induced void is found along the daisy chain. Voids are only localized at the cathode side in the feed line that is the metal line in the top die (Figure 7.14).

These results support the fact that an electron flow flowing perpendicularly to the hybrid bonding interface is favorable compared to a parallel one as, in this case, voids and extrusions can be observed in bonding interface [20]. In addition, one must notice that intrinsic bonding voids originating from interfacial copper oxide do not move with the electrical current showing that the hybrid bonding process is mature.

Complementary work also investigated the impact of seed layer type on EM, confirming the higher Cu/TaN/Ta adhesion energy and longer resistance to EM [16].

In conclusion, the hybrid bonding module has no impact on the EM resistance and presents excellent environmental reliability. The weakest spot is always the BEOL level (top or bottom depending on the electron flow).

(a)   (b)

FIGURE 7.14
Characterization of daisy-chain (100 connections) after electromigration tests. (a) Lock-in thermography results (amplitude, 15×, 0–0.15 V, 10 Hz, 60 seconds. White arrow locates a possible failure) and (b) FIB-SEM cross-section in the area indicated by the white arrow in (a).

More recently we have shown that final passivation annealing conditions could significantly influence the EM lifetime [21]. In the case of high-pressure Deuterium (HPD2) annealing, EM lifetime is reduced and the degradation is attributed to the presence of deuterium accumulated at the barriers and capping layers evidenced by physicochemical characterizations.

7.2.4 Hybrid Bonding Maturity Increase: Moving toward Production

Over the past years, the strong interest of industry for direct hybrid bonding has enabled to push forward this process through even more complex but realistic integrations with increasing number of metal levels. We have mentioned in the previous paragraph how this process can be sensitive to topology variation and that uniformity after CMP is essential to obtain perfect bonding pairs. Therefore, by increasing the number of underneath layers it can potentially degrade the final interfacial flatness.

Nevertheless, since 2015, several demonstrations have proven that the BEOL topology could be successfully managed to obtain perfect WtW bonding despite more than six BEOL levels in each wafer.

First work published in 2015 by L. Benaïssa et al. [5] reported the 200 mm wafer-level assembly of an advanced image sensor with control logic units and memories. The stack includes all back-end levels of a 0.13 µm double damascene. Thanks to a three-step CMP process applied at the bonding level and a proper wafer conditioning (control of each buried metal-level topology), the bonding was achieved at room temperature and under atmospheric pressure. Despite the total of 12 metal levels (5 + 5 BEOL and 2 added hybrid bonding levels), no defect was observed by SAM after final 400°C anneal (Figure 7.15) and alignment accuracy was measured at less than 400 nm. Excellent copper-to-copper connection has been observed and this final structure, which was the first demonstration of its kind at that time, confirms the potential of direct hybrid bonding.

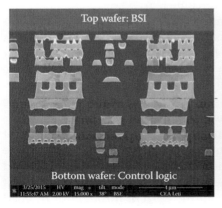 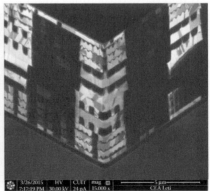

FIGURE 7.15
FIB SEM characterization of a 3D image sensor stack containing 12 metal levels.

Since then, SONY announced in March 2016 the production of the IMX260 3D Imager achieved, thanks to WtW hybrid bonding [23]. It is a 12MP camera constituted of a five-metal (Cu) CIS die and a seven-metal (6 Cu + 1 Al) image signal processor (ISP) die. The copper-to-copper pads are 3 µm wide and present a 14 µm pitch in the peripheral regions and a 6 µm pitch in the pixel array. This imager is already used as a rear-facing camera in the Samsung S7 mobile and confirms that WtW is mature for mass production.

The two examples mentioned are imager applications in which electrical signal exit is achieved thanks to wire bonding after top imager wafer thinning. However, to reduce the module size and to fasten data exchange with other chips, it has been recently shown that hybrid bonding can also be combined with a TSV–last process type [24].

7.2.5 Specificity of the Die-to-Wafer Process Variation

If, as said, the WtW hybrid bonding is already in production for the imager industry, it naturally raises some questions about yield.

Indeed, stacking two wafers together may erase all possible advantages of the integration, especially when considering insufficient yield from at least one of the two wafers stacked.

A DtW approach has then to be considered as an alternative. This approach has the advantage to stack known good dies (KGD) on known good dies. Consequently, a presumed 100% yield wafer may be *rebuilt*. Furthermore, the DtW integration would enable a multidie stacking with heterogenous functionality, which could widen the application field.

Nevertheless, additional difficulties of this integration lead, so far, to a reduced level of maturity: Proper handling of the dies is necessary, with a continuous compatibility with further copper–copper bonding: No degradation nor contamination of the top die surface after dicing is mandatory. However, in the work published by Y. Beillard et al., it has been demonstrated that these difficulties could be overcome [25].

In this published work, a complete comparison between WtW and DtW electrical performances has been achieved on 200mm wafers using daisy chains with connection number ranging from 4,872 to 29,422. The copper-to-copper area is $3 \times 3 \ \mu m^2$ and the pitch is ranging from 7 to 30 µm. In addition to prove that bonding interface is morphologically similar to whatever die- or wafer-level approach, the measured contact resistance was found to be identical to the one obtained in the WtW approach as well as the theoretical calculated value.

Nevertheless, if it is demonstrated now that pick-and-place equipment designed for hybrid bonding of DtW can have an alignment accuracy in the range of +/–1 µm (SET French Equipment company), then moving toward mass production is still challenging. The current hybrid bonding equipment throughput is very low, therefore far from the current throughput reached with standard mass reflow bonding using larger µbumps (>10,000 dies per hour). Nevertheless, we can predict that equipment suppliers may increase this throughput, a figure of 500 or 1000 dies by hour may be sufficient to make this integration economically viable, especially for very large dies.

To meet the DtW hybrid bonding throughput required by the industry, CEA-Leti has additionally developed since several years a self-assembly-based process, using capillary force between the top die and bottom wafer surfaces for aligning.

7.2.6 Die-to-Wafer Process Throughput Increase with Self-Assembly

Self-assembly process is based on the use of small water drops to align and bond dies to a wafer substrate. Chip alignment results from surface tension minimization of the water drop. Hybridation is achieved, thanks to direct bonding after the drop evaporation as presented in Figure 7.16.

A high-fluid containment is required to obtain a high accuracy in DtW self-assembly process. The aim of this containment is to avoid fluid overflowing beyond the die surface, which would lead to an alignment loss. The confinement area is obtained by physical and/or chemical contrast. The physical contrast consists of a topology modification at die edges. In this case, the water overflowing is controlled, thanks to the canthotaxis effect. Canthotaxis condition is detailed in Figure 7.17. The chemical contrast is created by the deposition of a hydrophobic material on die edges. It can be seen

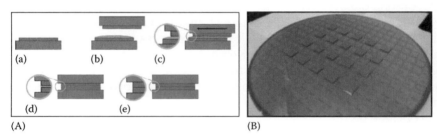

FIGURE 7.16
(A): Self-assembly process description: (a) bottom structure before stacking, (b) water drop deposition and top die prepositioning, (c) self-assembly thanks to capillary restoring forces, (d) die alignment, residual interfacial water film, and (e) water evaporation and dies direct bonding. (B): 200 mm electrical wafer stacked, thanks to self-assembly process with 20 top dies.

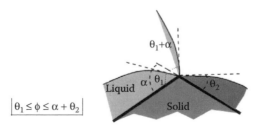

FIGURE 7.17
Principle and illustration of canthotaxis phenomena.

TABLE 7.2

Comparison between the Contact Resistances Obtained with Three Different Integration Processes

NIST Structure	Contact Resistance		
Hybridation Process	DtW (Self-Assembly)	DtW (Pick and Place)	WtW
Contact area (μm^2)	340×5	340×5	340×5
Copper resistivity ($\Omega \cdot \mu m$)	2.72×10^{-2}	2.06×10^{-2}	2.07×10^{-2}
Experimental resistance (Ω)	5.1	3.9	3.89
Theoretical resistance (Ω)	5.09	3.74	3.77
Variation (%)	**+0.19**	**+3.08**	**+4.1**

that the best theoretical confinement will be reached by creating both topology (high α value) and hydrophobic treatments (high θ_2 value).

If interesting demonstrations have already been made with μbumps assembly [26], CEA-Leti has adapted the technology to the constraints of hybrid bonding.

The process has been optimized, thanks to simple 1×1 cm nonpatterned dies with both topology and hydrophobic contrast on their edges. These structures enabled to demonstrate high alignment accuracy yield (>90%) and high bonding quality (below 1 μm) [27,28].

Finally, S. Mermoz et al. [29] demonstrated that the presence of interfacial water did not modify the electrical performances of the connected daisy chains (as shown in Table 7.2) for a Kelvin NIST structure. Furthermore, this work was continuously completed by theoretical simulation analysis using a numerical software to explain how outer perturbations could impact the capillary restoring force [30].

To conclude on this paragraph, hybrid bonding Cu/SiO$_2$ has gained many interest from industry players since few years, whereas first publications were released in early 2000. In between, impressive understanding studies of the bonding principle were carried out, many progresses about CMP and its material were achieved. Finally, production in a WtW scheme is now a reality for CIS and there is no doubt that other application fields will now continue the trend. DtW is the natural (but challenging) continuation of the roadmap.

7.3 3D Sequential: 3D Very-Large-Scale-Integration CoolCube™

To increase the density of layers' interconnects even more and to reach some pitches in the range of few tens of nanometers, an alternative technology to parallel stacking is needed. This paragraph will detail process and challenges of the FEOL-type high-density technology, the sequential integration CoolCube™.

7.3.1 3D Sequential: Principle

An alternative approach to conventional device scaling for future nodes is the 3D sequential integration (also referred as 3DVLSI (3D Very-Large-Scale-Integration) CoolCube™ or monolithic 3D integration) [31]. CoolCube™ process flow where devices are built one above the other in a sequential manner offers the possibility to stack devices with a lithographic alignment precision (few nm), enabling via density >100 million/mm² between transistors tiers (for 14 nm design rules). 3DVLSI can be routed either at gate or at transistor levels (Figure 7.18).

The partitioning at the gate level allows IC performance gain without resorting to scaling, thanks to wire length reduction [32]. Partitioning at the transistor level by stacking n-FET (field-effect transistor) over p-FET (or the opposite) enables the independent optimization of both types of transistors (customized implementation of performance boosters: channel material/substrate orientation/raised source and drain, strain, and so on. [31,33]) with reduced process complexity compared to a planar cointegration. This integration enables to merge several technologies together; for the bottom layer, it can be any CMOS from bulk planar to (fin field-effect transistor) FinFET or fully-depleted silicon-on-insulator (FDSOI). To benefit from the full 3D opportunities and avoid global routing congestion [33], there is a need to implement local routing of the bottom tier: Intertier metal layers need to be incorporated in the technology (Figure 7.19).

As a consequence, intermediate back-end-of-lines (iBEOL) levels need to support top FET thermal budgets. A reasonable maximum thermal budget for top FET has been determined around 500°C [34]. Indeed, the silicide has been identified as the main contributor for the bottom metal–oxide–semiconductor FET (MOSFET) instability. The summary of thermal limitation is shown in the graph given in Figure 7.20. The maximum top FET temperature is fixed around 500°C for 2 hours, however this temperature can clearly be increased when the anneal duration is reduced [35].

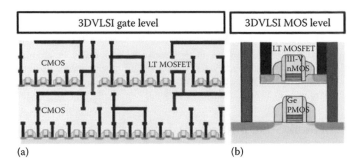

FIGURE 7.18
CoolCube™ integration: Partitioning schemes for 3DVLSI, namely (a) 3DVLSI at the gate level and (b) 3DVLSI at the transistor level.

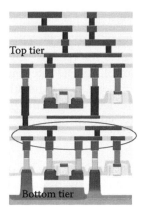

FIGURE 7.19
3DVLSI structure with two levels of intertiers interconnections.

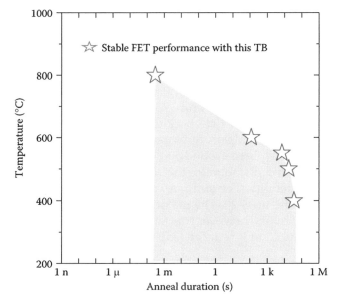

FIGURE 7.20
Summary of thermal budgets tested on FDSOI technology with SiGe channel for pFET and SiGe:B/SiC:P RSD with preserved N &PFET ION-IOFF performance. (From Fenouillet-Beranger, C. et al., New insights on bottom layer thermal stability and laser annealing promises for high performance 3D VLSI, *IEEE IEDM*, 2014.)

7.3.2 3D Sequential: State of the Art

Depending on the partitioning scheme and the technologies used for bottom and top layers, CoolCube™ addresses challenges associated to CMOS scaling and to More than Moore applications. One of the main difference in the 3D sequential integration demonstrations is the top silicon layer formation method

(Seed window [36], Poly-Si and laser-crystallized Epi-like Si [39] or oxide direct bonding [32]). Direct bonding clearly stands out with respect to the other techniques: Thanks to the high crystalline quality of top silicon layer, the devices' performance outperforms the other ones. Regarding CMOS scaling and especially SRAM integration, there have been great demands for higher density in all areas of SRAM applications such as network and cache standalone memory, and embedded memory of the logic devices. 3DVLSI integration is very promising for this type of applications as evidenced by the number of publications in literature [3,33,36–39]. Here again, as evoked previously, the best transistor performances are always achieved in case of top silicon crystalline layer obtained by oxide wafer bonding [3,31,33]. The ultimate example of high-performance CMOS at low process cost is the stacking of III–V nFETs above SiGe pFETs [40,41]. These high-mobility transistors are well suited for 3DVLSI because their process temperatures are intrinsically low. 3D sequential integration, with its high contact density, can also be seen as a powerful solution for heterogeneous cointegrations requiring high 3D via densities such as computation immersed in memory [42], heterogeneous IoT (Internet Of Things) chip [43,44], nano-electromechanical systems (NEMS) with CMOS for gas-sensing applications [45,46] or highly miniaturized imagers [47].

7.3.3 3D Sequential: Integration Process Flow and Low-Temperature Top FETs

The process flow principle is detailed in Figure 7.21 [48].

Bottom layer is firstly processed up to a few metal layers within a standard CMOS flow. In case of thin-film integration, a blanket wafer is subsequently transferred at low temperature on top of the bottom layer. It is realized through direct bonding either by a silicon-on-insulator (SOI) wafer [49] or by using

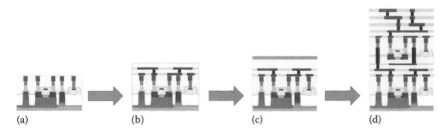

(a)   (b)   (c)   (d)

FIGURE 7.21
Process flow principle of monolithic 3D integration. By resorting to a unique alignment flow throughout the whole process, layers are stacked on top of each other within a lithographic alignment precision. (a) Bottom-layer processing with plugs down to the CMOS, (b) fabrication of the inter level lines to ensure short distance connection with bottom layer, (c) high-quality top film transfer by direct oxide–oxide bonding, and (d) top-layer fabrication and connection. (From Vinet, M. et al., Monolithic 3D Integration: A powerful alternative to classical 2D scaling, *IEEE* S3S, 2014. © 2014 IEEE. With permission.)

hydrogen implantation and splitting [50]. Top-layer lithography relies on a single-stream alignment flow, leading to the excellent alignment precision between top and bottom transistors. As mentioned before the top devices' process flow is limited to 500°C for a couple of hours to not degrade the bottom one's performance. The temperature of four key process modules needs to be lowered to match the thermal budget specifications to ensure bottom MOSFET and interconnects preservation. Analysis of the top MOSFET process flow shows that the dopant activation is the most critical thermal budget contributor. Indeed, in a standard high-temperature process Rapid Thermal Annealing (RTA), activation anneal steps are performed above 1000°C with a time scale around 1 second. The three other process modules to address are the offset spacers (~630°C), the raised source drain epitaxy (SiGe 30% at 650°C or Si at 750°C), and the gate-stack stabilization. Solid-Phase-Epitaxy-Regrowth (SPER) at 600°C combined with extension first implantation for dopant activation yields FDSOI devices with similar ION/IOFF performances than devices activated at standard temperatures [51]. Nanosecond laser anneal with some nanoseconds (ns) pulse duration (with low in depth thermal diffusion) is effective to activate dopants in FDSOI devices with sheet resistance below the high-temperature reference [52]. Finally, low-k spacers (for parasitic capacitance reduction) with lower thermal budgets than for SiN atomic layer deposition (ALD) liners could be used (SiCO) [53]. The epitaxy step has been successfully demonstrated at 500°C [54] by using cyclic deposition and etch process. Regarding the gate stack, the reliability has been demonstrated at 525°C [55] but further studies toward decreasing temperature while preserving reliability are needed especially in case of P-type metal–oxide–semiconductor (PMOS) devices. In order to alleviate this problem and then to relax the maximum thermal budget allowed for top-level processing in 3DVLSI integration, bottom transistors' silicide stability must be increased. Thanks to process boosters (introduction of silicon capping and pre-amorphizing implantation [PAI]), the thermal stability of NiPt can be extended [56]). In parallel the introduction of a novel silicide, $Ni_{0.9}Co_{0.1}$, seems very promising [57]. Thus, the main changes with respect to standard fabrication flow are summarized in Figure 7.22, highlighting that, thanks to the bottom silicide stability improvement, all different high-temperature modules have been optimized to fulfill the targeted temperature criteria for 3D sequential integration [58].

Focusing on contamination issue, all the CoolCube™ process should be realized within an ultraclean environment respecting stringent contamination constraints imposed by an industrial clean room. The different species can be monitored along the process through different steps, especially on the bevel edges, using Vapour Phase Decomposition (VPD), Droplet Collection (DC), and analysis by Inductively Coupled Plasma Mass Spectrometry (ICPMS). An example of results is summarized in Table 7.3 and confirms the very low Ni concentration fully compatible with a front-end environment demonstrating the efficiency of the contaminants' containment strategy developed at CEA-Leti [3].

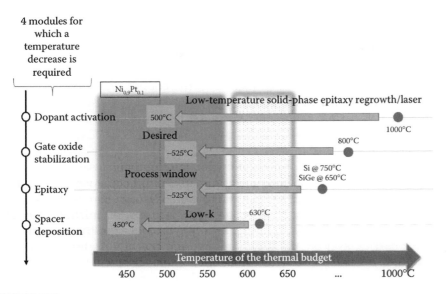

FIGURE 7.22
Bottom-layer thermal stability has been increased, thanks to silicide process optimization, (From Vinet, M. et al., *IEEE ESSDERC*, 2016.) and top-layer thermal budget has been decreased by optimization of hot-temperature modules.

TABLE 7.3

VPD-DC-ICPMS Monitoring of Ni through the Process on the Bevel. Low Limit of Detection = 6.5E7 at/cm²

	Ni (at/cm²)
Cleaning before bonding	<9.4E8
Bonding annealing	<9.4E8
High-k deposition	2.9E9
Gate stack etch	8.3E9
Epitaxy	6.4E+09
Dopant activation annealing	2.8E9

7.3.4 3D Sequential: Intermediate Back-End-of-Line

As for the bottom devices, iBEOL levels need to endure top FET thermal budgets implying to find solutions for implementing back-end material stable beyond 400°C. Currently the combination of copper with ultra low-k (ULK) materials is widely used for standard BEOL (low resistivity and capacitance, and thus speed improvement). However, the integration of such materials in the iBEOL of a CoolCube™ integration faces a number of challenges. Indeed, copper metallization can cause contamination issues in the case of wafer break during the process of the top transistor where FEOL contamination environment is required. An alternative to copper can be the use of tungsten (W) as it has already been integrated in the FEOL of several products. CEA-Leti studies [59] highlighted that Cu and W interconnections combined with ULK are stable up to 500°C for 2 hours and 550°C for 5 hours, respectively, in case of line 1 integration with 28 nm design rules. However, W resistance is still larger as compared to the copper one (by a factor 6). Moreover, as presented in the literature, the ULK stability is still questioned beyond 500°C [60]. Indeed, the modification of ULK structure and permittivity during a thermal anneal at temperature higher than 500°C may increase the leakage and delays of the iBEOL and thus may degrade the circuit performances. Thermal stability of dielectrics has been studied to select the most appropriate ones. Depending on the thermal budget set by the top MOS layer, various couple of materials are possible and summarized in Table 7.4 [61]. Regarding the barrier layer, SiCO seems to be the most promising material due to its robust composition, low permittivity (4.5) (lower than the state-of-the-art barrier layer SiCNH (5.6)), and its high thermal stability. However, for a top thermal budget limited to 500°C the state of the art SiCNH is still suitable. Regarding the oxide-based material, several of them seem suitable depending on the top feasible thermal budgets. For a thermal budget limited to 500°C, 2 hours, the state-of-the art ULK (2.5) material is still possible. On the other hand, at 600°C, 2 hours, only the SiO_2 is suitable. Finally, the permittivity of these materials is crucial to avoid circuit performance degradations.

The reliability of ULK dielectrics after a relatively high thermal budget has never been demonstrated up to now. Using the standard extrapolation parameters, the extrapolated lifetime is extracted for W/ULK interconnections [55] (Table 7.5). Although time to failure (TTF) decreases with increasing thermal budget, a lifetime (line to line) of 10^9 years is calculated for the highest thermal budget, larger as to Cu/ULK intermetal dielectric (IMD) lifetime

TABLE 7.4

Possible Dielectrics for the iBEOL as a Function of the Top FET Thermal Budget and Associated Permittivity

	State-of-the-Art	500°C, 2 hours	550°C, 2 hours	600°C, 2 hours
Barrier layer	SiCNH (5.6)	SiCO (4.5)/SiCN(5.6)	SiCO (4.5)	SiCO (4.5)
Oxide-based	ULK (2.5)	ULK (2.5)	Dense LK (3)	SiO_2 (4.1)

Source: Xu, H. et al., *Design, Automation & Test in Europe (DATE)*, 1–6, 2011.

TABLE 7.5

Extrapolated Lifetime (TTF) of Intermetal Diel ULK with Ti/TiN/W before and after Anneals. Years Needed to Reach the Failure Rate (<1%) with an Operating Voltage at 1.115 V

	No Anneal	500°C	550°C
Ti/TiN/W (years)	4.10^{15}	(5 hours) 5.10^{11}	(5 hours)10^{09}

Source: C-M. V. Lu et al., Key process steps for high performance and reliable 3D sequential integration, *IEEE VLSI* 2017.

without thermal budget. Therefore, W/ULK reliability is not an issue for the 3D sequential integration due to the high initial lifetime, as top transistor thermal budget is limited to 500°C for a couple of hours.

In spite of contamination issue, it could be very interesting to integrate Cu/ULK lines in the iBEOL of 3D sequential integration to use standard bottom tier process, however the reliability versus thermal budget should be studied.

Integration of dense W lines introduces a new complexity degree not only in terms of contamination but also in terms of CMP. Indeed, in spite of the high metal lines density, the planarity of the structure should be preserved to ensure a good bonding quality. A layer transfer by wafer bonding has been realized above a 28 nm industrial metal 1 short loop with line densities up to 70% [3]. The schematic of the experiment is illustrated in Figure 7.23. W filling is used instead of standard copper, coupled with a Ti/TiN barrier. A high-quality layer transfer as observed with acoustic microscopy observation is presented on Figure 7.24. Figure 7.24a shows a SiO_2/SiO_2 reference bonding (without any defect that would appear as white dot), whereas Figure 7.24b shows the bonding above W lines with very few bonding defects at the wafer edges. These residual defects are explained by a nonfully optimized W lines CMP process that will be easily adjusted on a real lot. The bonded structure was then thinned down to the top buried oxide (BOX) with both grinding and tetramethylammonium hydroxide (TMAH) etching. A SEM cross-section of the structure after thinning with the Si layer highlighted in red is presented in Figure 7.24c.

7.3.5 3D Sequential: 300 mm Electrical Demonstration

For the first time in 2016 [3], a full CMOS over CMOS 3D sequential integration has been demonstrated with a top level compatible with state of the art high-performance FDSOI process requirements such as high-k/metal gate

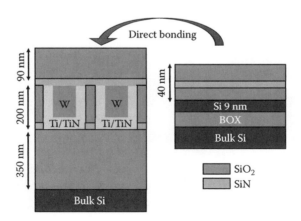

FIGURE 7.23
Description of the Si layer transfer above W metal 1 level. Ti/TiN diffusion barrier is used.

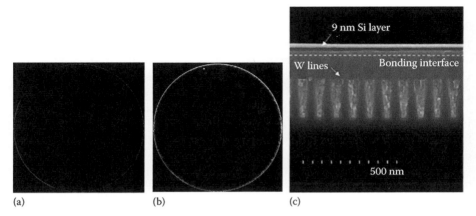

FIGURE 7.24
(a) Acoustic microscopy on a reference SiO_2/SiO_2 bonding. Perfect bonding is observed (no defect) and (b) acoustic microscopy on the studied structure. High-quality bonding reached. Some defects on the edge due to W over-polishing. (c) SEM cross-section of the bonded structure, with bottom W lines, after thinning. The 9 nm Si layer is highlighted in dark gray.

and raised source and drain. On the bottom level, standard high-temperature NMOS is N-type metal–oxide–semiconductor (NMOS) and PMOS transistors with HfSiON/TiN gate stack are fabricated on 300 mm SOI wafers (t_{BOX} = 145 nm/t_{Si} = 7 nm) with Si raised source and drain junctions activated at 1050°C. For the top level, a HfO_2/TiN gate stack was formed followed by a spacer zero deposition at 630°C and a selective $SiGe_{27\%}$ raised source drain epitaxy at 650°C for both NMOS and PMOS transistors. The junctions were activated by SPER technique during 1 minute at 600°C. Figure 7.25 shows the TEM cross section of the 3D sequential contacted structure with a focus on two stacked transistors. The Id-V_G characteristics for both NMOS and PMOS

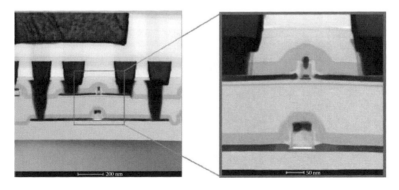

FIGURE 7.25
TEM cross-section of the 3D sequential structure.

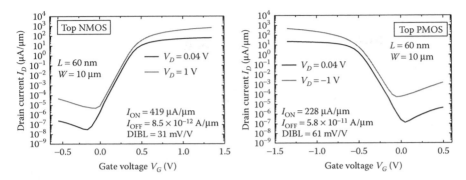

FIGURE 7.26
ID-VG characteristics of both top-level NMOS and PMOS transistors. $W = 10$ μm, $L_g = 60$ nm. (From Brunet, L. et al., First demonstration of CMOS over CMOS 3D VLSI CoolCube™ integration on 300 mm wafers, *IEEE VLSI*, 2016. © 2016 IEEE. With permission.)

transistors on the top level are presented in Figure 7.26. Finally, for the first time the voltage transfer characteristics of 3D sequential inverters with either NMOS or PMOS on the top level, on 300 mm wafer, are presented in Figure 7.27.

As a conclusion, 3DVLSI sequential 3D integration CoolCube™ successfully achieved in the past 10 years to prove the feasibility of the top layer manufacturing and its robustness. This innovative integration, in a blaze of many high-level worldwide publications, could be the answer for the end of Moore's law that some specialists predicted in a near future. In addition, in the same manner as for advanced packaging or 3D high density, heterogeneous integration may be also a key advantage offered by CoolCube™, in a range of pitch that is more aggressive than parallel stacking.

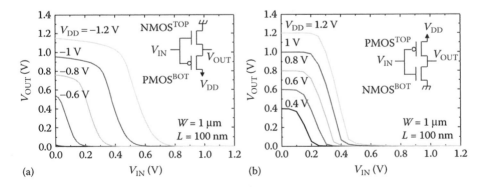

FIGURE 7.27
(a) Voltage transfer characteristics of a 3D sequential inverter with NMOSFET on top and PMOSFET on the bottom level. (b) Voltage transfer characteristics of a 3D sequential inverter with PMOSFET on top and NMOSFET on the bottom level.

7.4 3D Technologies Comparative Thermal Analysis

One the main challenge of 3D technologies, specifically high-density 3D integration, is power density and associated thermal dissipation. This is often put ahead as a potential strong show stopper. Indeed, due to the increased number of die layers, a too large power density may arise, that would imply a large thermal flux, which might be difficult to dissipate using standard packaging technologies. In an extreme manner, as a result, thermal runaway would arise that would potentially destroy the 3D circuits.

The thermal dissipation of commonly used 3D-packaging technologies using TSV has already been widely studied [61–64]. For instance, at CEA-Leti, a systematic thermal exploration [64] has been carried out to investigate the key differentiating parameters of TSV-based 3D ICs: power density, die thinning, die-to-die thermal coupling, and the impact of TSVs. The study reveals that nonthinned dies in a 3D stack may act as heat spreaders, whereas TSVs may even provoke exacerbated hotpots due to the dielectric layer used for TSV isolation. TSVs induce a decreased horizontal Heat Transfer Coefficient (HTC) cause of the dielectric layer for a light increase in the vertical heat path. TSVs lead to larger thermal hotspots in case of lateral dissipation blockage, contrarily to the common believes [65]. Measurements on a dedicated test-chip and systematic simulations have been carried out leading to the same conclusions [64,66].

Nevertheless, little work has been carried out on thermal study of denser 3D technologies such as the ones described in the previous paragraphs [67]. In this section a thermal comparative study is carried out between the respective new advanced 3D technologies presented in the above-mentioned sections: Cu/SiO$_2$ hybrid bonding and CoolCube™ [68]. This comparative thermal study between standard μbumps, Cu/SiO$_2$ hybrid bonding, and CoolCube™ is performed for multiple-die stacks and considering a full range of technology parameters and different application scenarios.

Section 7.4.1 presents the respective technology parameters to setup a thermal comparative study and the associated applicative scenarios, whereas the following one presents the associated thermal results.

7.4.1 3D Technology Parameters for Thermal Comparison

CEA-Leti has monitored a comparative thermal study to compare three different 3D technologies: two parallel 3D technologies (TSV based with μbumps or with Cu/SiO$_2$ bonding) and the sequential approach CoolCube™, for a different number of parameters: number of active layers, die thickness, interconnect density, and power scenarios. In order to allow fair comparison, as a hard constraint the total die thickness and package are kept fixed. Figure 7.28 shows an example of stacking diagram used for thermal simulations and indicates circuit and 3D technology parameters, details are

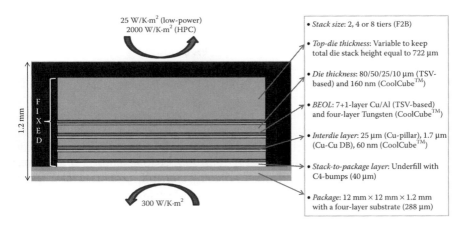

FIGURE 7.28
Example of stacking diagram indicating the main circuit and 3D integration technology parameters considered for thermal simulations.

TABLE 7.6

3D Technology Parameters: Material and Thickness

| Layer | Parallel 3D Technologies | | | | Sequential 3D Technology | |
| | μbumps | | Cu/SiO$_2$ Direct Bonding | | CoolCube™ | |
	Material	Thickness	Material	Thickness	Material	Thickness
BEOL	Cu/Al+SiO$_2$	7+1 layers 7 μm	Cu/Al+SiO$_2$	7+1 layers 7 μm	W+SiO$_2$	4 layers[a] 1.72 μm
Interdie layer	Cu/SnAg+ polymer underfill	25 μm	Cu+SiO$_2$	1.7 μm	W+SiO$_2$	60 nm
Die[b] substrate	Si	80/50/25 μm	Si	50/20/10 μm	Mostly SiO$_2$	160 nm

[a] Except for bottommost tier, which has a standard 7+1 Cu/Al BEOL configuration.
[b] Except the topmost tier, which is as thick as the die stack height limit permits.

given in Table 7.6. The 3D stack contains 1–8 tiers, which are assembled in a face-to-back (F2B) manner [69]. In the case of TSV-based integration, the thickness of the die substrate spans from 25 to 80 μm for assemblies using μbumps and from 10 to 50 μm for Cu/SiO$_2$ hybrid bonding. The layout of each tier is based on a real scalable 65 nm 3D circuit where identical tiers are stacked one on top of each other. An exception is made for CoolCube™, which uses tungsten for the metallization layers (BEOL) of the intermediate tiers due to thermal limitations and contamination risks during the fabrications process, as explained in the previous paragraph. The height and diameter of the μbumps in the interdie layer are, respectively, 25 and 20 μm, thus in the typical range for this technology. The thickness of the interdie layer is 1.7 µm for the Cu/SiO$_2$ hybrid bonding and only 60 nm in the case of CoolCube™. As already mentioned, the impact of the TSVs on the thermal conductivity of the silicon substrate has been demonstrated to be very limited [65] and therefore will not be considered in this study.

The 3D die is stacked on a flip-chip organic package with the bottom tier connected to a four-layer ball grid array (BGA) package (288 µm height) through C4-bumps (40 µm height). To be compatible with typical low-power applications, such as mobile, the total package height was fixed at 1.2 mm, thus resulting in a max height of 722 µm for the 3D die stack. Provided this constraint, the topmost tier is made as thick as possible as it plays as a heat spreader for hot-spot mitigation. Specific HTCs were applied to the boundary surfaces according to the considered power application scenario to emulate the behavior of typical Printed Circuit Board (PCB) and package components such as Thermal Interface Material (TIM) heat spreader and heat sink.

Figure 7.29 illustrates the power profiles used to emulate multiple application scenarios and to provoke distinct thermal behaviors depending on the application use case. Heat dissipation at hot spots is primarily diffused through the silicon substrate and usually spreads in a semispherical direction, rapidly decreasing the heat density and lowering the peak temperature. But for 3D ICs, the thinned silicon substrate reduces the lateral heat spreading capability and the interdie layers (mainly polymer) act as vertical thermal barriers, which results in exacerbated hot spots. When power is evenly distributed, the temperature gradient over the tier is much smaller and the heat flows mostly perpendicular to the die substrates in the 3D stack.

Most of the heat generated in applications with intensive power dissipation, such as HPC, flows toward efficient heat sinks usually mounted on top of the package. On the contrary, for applications with limited power dissipation, circuit is mainly cooled from the bottom and heat flows toward the PCB through the package substrate.

This study also considers applications with similar power budget allocated either in a single or in multiple tiers. Taking into account voltage and

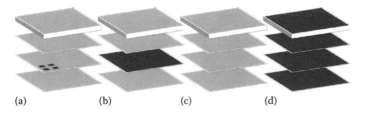

FIGURE 7.29
Examples of power application scenarios. (a) Strong hot spots in a single tier, (b) limited uniform power in a single tier, (c) limited uniform power distributed between tiers, and (d) intensive uniform power dissipation.

TABLE 7.7

Power Dissipation Profiles, according to Application Scenarios Described in Figure 7.29

	Hot Spot		Single-Tier		Multitiers		Multitiers HPC	
Quantity of Tiers	Total Power (W)	Power Density[a] (W/cm²)	Total Power (W)	Power Density[a] (W/cm²)	Total Power (W)	Power Density[a] (W/cm²)	Total Power (W)	Power Density[a] (W/cm²)
2	2.64	365	2.657	3.65	2.657	1.825	5.3	3.65
4	2.64	365	2.657	3.65	2.657	0.913	10.6	3.65
8	2.64	365	2.657	3.65	2.657	0.456	21.2	3.65

[a] In active regions.

frequency scaling, to distribute tasks in multiple CPUs running at a lower frequency may result in higher throughput for same power consumption than running in a single CPU at high frequency [70]. Total power dissipation and power density for the different applications scenarios in Figure 7.29 are listed in Table 7.7.

Thermal modeling and simulation for this large set of technology parameters and power applications have been carried out using a thermal analysis prototype provided by Mentor Graphics Calibre® [71]. This powerful thermal analysis solution, which relies on the FloTHERM simulation engine, has an automated effective thermal property extraction (EFFP) feature that allows to easily account for complex fine-grain structures such as the BEOL, interlayer vias, and µbumps arrays used for 3D connections.

7.4.2 Comparative Study Thermal Results

This section brings thermal simulation results for the Cu/SiO_2 hybrid bonding, CoolCube™, and µbumps technologies with regard to die thickness, die-to-die connectivity density, and the number of stacked tiers. Due to their very distinct thermal behaviors, hot spot and uniform power application profiles are investigated in separate subsections. The thermal conductivity of the interdie layer is highly dependent on the number of die-to-die 3D connections, more precisely the ratio of metal and insulator within this layer. This study considers two distinct corners for this metal density parameter either (1) very few die-to-die connections (ratio <1%) or (2) a high die-to-die connectivity density (ratio of 16% for CoolCube™ and ratio of 25% for µbumps and hybrid direct bonding). In most real circuits, the actual die-to-die connectivity density will lie in between these two corners due to design and layout constraints.

7.4.2.1 For Hot-Spot Dissipation Scenarios

Figure 7.30 brings the peak temperature results for hot-spot dissipation scenarios (case [a] in Figure 7.29). The interdie interface layer and die thickness play important roles as they directly affect the thermal coupling between adjacent tiers and the lateral heat dissipation, respectively. CoolCube™ presents very tight thermal coupling and heat can easily reach the top tier to benefit from its thickness to spread out. For μbump assembly technology, the underfill (nonconductive polymer, which mechanically bonds the dies together) layer acts as a strong thermal barrier between the hot spots and the heat sinks and provoke higher peak temperatures. Cu/SiO_2 hybrid bonding

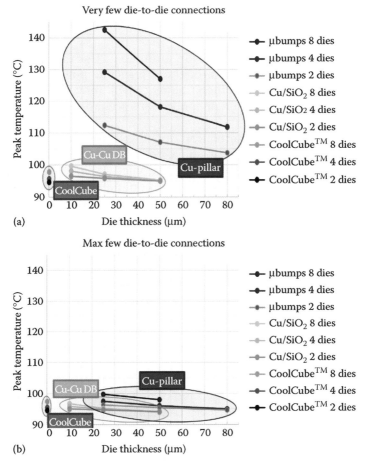

FIGURE 7.30

(a) Peak temperature in the case of hot-spot dissipation considering very few die-to-die connections (ratio <1%) or (b) max connection density between tiers (ratio of 25% for TSV-based and 10% for CoolCube™). Hot spots are active only in midmost tier to account for similar heat flow paths in both vertical directions.

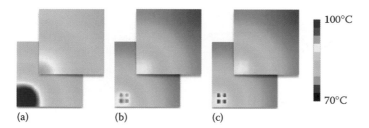

FIGURE 7.31
Thermal maps of the middle and topmost tiers in the case of hot-spot dissipation (Figure 7.29a) in a eight-dies stack: (a) parallel stacking with μbumps, (b) parallel stacking with Cu/SiO$_2$ direct bonding, and (c) sequential stacking CoolCube™.

benefits from both thermal coupling and lateral heat dissipation in the intermediate tiers, therefore presenting the best thermal performance in most 3D stacks with more tiers. This difference in the thermal profiles is depicted in the temperature maps in Figure 7.31, where the hot spot outlines reveal the lateral heat dissipation through the die substrates.

7.4.2.2 For Uniformly Distributed Power Dissipation Scenarios

Figure 7.32a and b show results for uniform power dissipation scenarios where the different 3D technologies have similar peak temperatures. In this case, most of the heat flows vertically and the 3D technology parameters have reduced thermal impact as external components such as the package structures dominate the vertical thermal resistance of the circuit. In case the total power dissipation allocated in a single tier or distributed in multiple tiers leads to about the same temperature results for most of the 3D configurations. This brings a perspective of increasing the processing capability of the circuit by allocating tasks into different tiers of the 3D stack while keeping the same power consumption and preserving the thermal performance. In the case where every additional stacked tier results in extra power dissipation, and therefore the resulting increase in the peak temperature is nearly linear with regard to the number of stacked tiers.

7.4.3 Thermal Comparison Conclusion

The study presented in this section reports a comparison of the thermal performance of the CoolCube™ and Cu/SiO$_2$ hybrid bonding technologies versus widely used μbump technology. It highlights the contribution of individual technology parameters to the overall thermal performance of the 3D stacks considering various power application scenarios. For the sake of comparison, the respective technologies are compared for different scenarios and for fixed packaging. As conclusion, the μbump technology shows poor results for thermal hot spots, due to underfill layers that have poor thermal

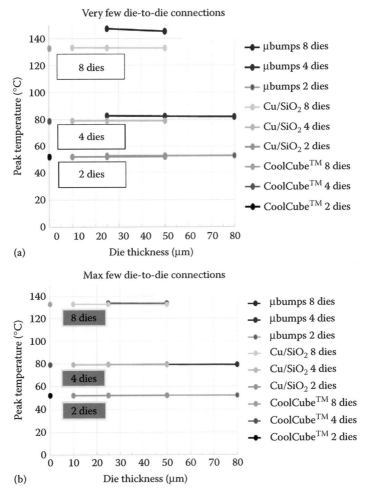

FIGURE 7.32
Peak temperature resulting from a power application scenario with intensive power dissipation uniformly distributed over the tiers (Figure 7.29d) and in the case of (a) very few connections between tiers and (b) max density connections between tiers.

conductivity. On the other hand, Cu/SiO_2 hybrid bonding and CoolCube™ technologies present a very good thermal coupling between the different circuit layers, where the impact of interconnect density is rather limited. In the case of uniform power dissipation, the thermal result is more or less linear with power density, then with the number of active dies. It is known that 3D integration offers the possibility of running circuits at different performance modes [70], where some of them are not practical in 2D.

7.5 General Conclusion

3D high-density integrations (typically below a 10 μm pitch of interconnects) may require some alternative technologies to widely used μbumps. In this section, CEA-Leti presents two alternative but complementary approaches. The first one is similar to a BEOL approach, using Cu/SiO_2 hybrid bonding well adapted for a pitch of interconnect superior to 1 μm. The second one is nearer to FEOL processes, named CoolCube™, well adapted for pitch below 1 μm. This chapter gives all details on technology and results, proving that those alternatives offer high performances without the need of More than Moore advanced nodes. Meantime, Cu/SiO_2 hybrid bonding and CoolCube™ have the potential to solve the interconnection density limitation of the TSV-based integration processes. Furthermore, despite active devices being integrated in extremely thin layers, and contrarily to common believes, Cu/SiO_2 hybrid bonding and CoolCube™ technology process exhibit thermal performance similar to or even better than that of the 3D integration using μbumps and associated underfill considering the same power applications. The very tight die-to-die thermal coupling allows easy heat dissipation of internal hot spots without any noticeable temperature difference between the stacked layers.

References

1. L. Di Cioccio et al., An overview of patterned metal/dielectric surface bonding: Mechanism, alignment and characterization, *JECS*, 2011, pp. 81–86.
2. M. Goto et al., Three-dimensional integrated CMOS image sensors with pixel-Parallel A/D converters fabricated by direct bonding of SOI layers, *IEEE, Electron Device Meeting IEDM*, 2014.
3. L. Brunet et al., First demonstration of CMOS over CMOS 3D VLSI CoolCube™ integration on 300 mm wafers, *IEEE VLSI*, 2016.
4. V. Balan et al., CMP process optimization for bonding applications, *ICPT 2012*, October 15–17, 2012, Grenoble, France, pp. 177–183.
5. L. Benaissa et al., Next generation image sensor via direct hybrid bonding, *2015 IEEE 17th Electronics Packaging Technology Conference*, December 2–4, 2015, Singapore.
6. B. Rebhan et al., <200 nm wafer-to-wafer overlay accuracy in wafer level Cu/SiO2 hybrid bonding for BSI CIS, *Electronics Packaging Technology Conference*, December 2–4, 2015, Singapore.
7. M. Okada et al., High-precision wafer-level Cu-Cu bonding for 3DICs, *IEEE, Electron Device Meeting IEDM*, 2014.

8. S. Lhostis et al., Reliable 300 mm wafer level hybrid bonding for 3D stacked CMOS image sensors, *2016 IEEE 66th Electronic Components and Technology Conference*, May 31–June 5, 2016, Las Vegas.

9. L. Wang et al., Direct bond interconnect (DBI) for fine-pitch bonding in 3D and 2.5D integrated circuits, *Pan Pacific Microelectronics Symposium*, 2017, IEEE.

10. Z.Y. Liu et al., Detection and formation mechanism of micro-defects in ultrafine pitch Cu-Cu direct bonding, *Chinese Physics B*, 25(1), 018103, 2016.

11. Q.Y. Tong and U. Gösele, *Semiconductor Wafer Bonding Science and Technology*, 1999, Wiley, New York, 320 p.

12. W.P. Maszara et al., Bonding of silicon wafers for silicon–on-insulator, *Journal of Applied Physics*, 64(10), 4943–4950, 1988.

13. P. Gueguen et al., Copper direct-bonding characterization and its interests for 3D integration, *Journal of the Electrochemical Society*, 156(10), H772–H776, 2009.

14. P. Gueguen et al., Direct bonding: An innovative 3D interconnect, *ECTC Proceeding*, 2010, pp. 878–883.

15. L. Di Cioccio, F. Baudin, P. Gergaud et al., Modelling and integration phenomena of metal-metal direct bonding technology, *ECS Transactions*, 64(5), 339–355, 2014.

16. Y. Beilliard et al., Advances toward reliable high density Cu–Cu interconnects by Cu-SiO2 direct hybrid bonding, *IEEE International 3D System Integration Conference (3DIC)*, 2014.

17. Imec and EVG demonstrate for the first time 1.8 μm pitch overlay accuracy for wafer bonding. https://www.evgroup.com/en/about/news/2017_01_imec.

18. R. Taïbi et al., Full characterization of Cu/Cu direct bonding for 3D integration, *ECTC Proceedings*, 2010, pp. 219–225.

19. R. Taïbi et al., Investigation of stress induced voiding and electromigration phenomena on direct copper bonding interconnects for 3D integration, *IEEE Electron Device Meeting IEDM*, 2011.

20. S. Moreau et al., Mass transport-induced failure of hybrid bonding-based integration for advanced image sensor applications, *2016 IEEE 66th Electronic Components and Technology Conference*, May 31–June 5, 2016, Las Vegas, NV.

21. J. Jourdon et al., Effect of pasivation annealing on the electromigration properties of hybrid bonding stack, *IEEE International Reliability Physics Symposium (IRPS)*, 2017.

22. A. Garnier et al., Electrical performance of high density 10 μm diameter 20 μm pitch Cu-Pillar with chip to wafer assembly, *IEEE 67th Electronic Components and Technology Conference*, May 30–June 2, 2017, Orlando.

23. Sony IMX260 in Samsung Galaxy S7: Stacked or not?. http://image-sensors-world.blogspot.fr/2016/03/sony-imx260-in-samsung-galaxy-s7.html.

24. C. Cavaco et al., Copper oxide direct bonding of 200 mm CMOS wafers with five metal levels and TSVs: Morphological and electrical characterization, *JECS*, 2016, pp. 43–46.

25. Y. Beillard et al., Chip to wafer copper direct bonding electrical characterization and thermal cycling, *IEEE International Conference on 3D System Integration (3D IC)*, 2013.

26. T. Fukushima et al., Transfer and non-transfer stacking technologies based on chip-to wafer self-assembly for high throughput and high precision alignment and microbumps bonding, *IEEE International 3D Systems Integration Conference*, 2015.

27. L. Sanchez et al., Chip to wafer direct bonding technologies for high density 3D integration, *IEEE Electronic Components and Technology Conference*, 2012.

28. S. Mermoz et al., Impact of containment and deposition method on sub-micron chip-to-wafer self-assembly yield, *IEEE International Conference on 3D System Integration (3D IC)*, 2012.
29. S. Mermoz et al., High density chip-to-wafer integration using self-assembly on the performances of directly interconnected structures made by direct copper/oxide bonding, *IEEE 15th Electronics Packaging Technology Conference*, December 11–13, 2013.
30. J. Berthier et al., Self-alignment of silicon chips on wafer: The effect of spreading and wetting, *Sensor & Transducers*, 2012.
31. P. Batude et al., Advances in 3D CMOS sequential integration, *IEEE IEDM*, 2009.
32. Y.-J. Lee, P. Morrow, and S. K. Lim, Ultra high density logic designs using transistor-level monolithic 3D integration, *International Conference on Computer-Aided Design*, 2012.
33. P. Batude et al., Demonstration of low temperature 3D sequential FDSOI integration down to 50 nm gate length, *IEEE VLSI*, 2011.
34. C. Fenouillet-Beranger et al., New insights on bottom layer thermal stability and laser annealing promises for high performance 3D VLSI, *IEEE IEDM*, 2014.
35. P. Batude et al., 3DVLSI with CoolCube process: An alternative path to scaling, *IEEE VLSI*, 2015.
36. S.-M. Jung et al., High speed and highly cost effective 72M bit density S^3 SRAM technology with doubly stacked Si layers, peripheral only CoSix layers and tungsten shunt W/L scheme for standalone and embedded memory, *IEEE VLSI*, 2007.
37. C.C. Yang et al., Enabling low power BEOL compatible monolithic 3D nanoelectronics for IoTs using local and selective far infrared ray laser anneal technology, *IEEE IEDM*, 2015.
38. T. Naito et al., Worlds first monolithic 3D FPGA with TFT SRAM over 90 nm 9 layer Cu CMOS, *IEEE VLSI*, 2010.
39. T.T. Wu et al., Sub 50nm monolithic 3D IC with low power CMOS inverter and 6T SRAM, *IEEE VLSI-TSA*, 2015.
40. T. Irisawa et al., Demonstration of ultimate CMOS based on 3D stacked InGaAs-OI/SGOI wire channel MOSFETs with independent back gate, *IEEE VLSI*, 2014.
41. V. Deshpande et al., Advanced 3D monolithic hybrid CMOS with sub 50nm gate inverters featuring replacement metal gate (RMG)–InGaAs nFETs on SiGe-OI fin pFETs, *IEEE IEDM*, 2015.
42. Max M. Shulaker et al., Monolithic 3D integration of logic and memory: Carbon nanotube FETs, resistive RAM, and silicon FETs, *IEEE IEDM*, 2014.
43. Tsung-Ta Wu et al., Low cost and TSV free monolithic 3D-IC with heterogeneous integration of logic, memory and sensor analogy circuitry for Internet of Things, *IEEE IEDM*, 2015.
44. Fu-Kuo Hsueh et al., First fully functionalized monolithic 3D+ IoT chip with 0.5V light-electricity power management, 6.8 GHz wireless-communication VCO, and 4-layer vertical ReRAM, *IEEE IEDM*, 2016.
45. I. Ouerghi et al., Polysilicon nanowire NEMS fabricated at low temperature for above IC NEMS mass sensing applications, *IEEE IEDM*, 2014.
46. P. Batude et al., 3-D Sequential Integration: A key enabling technology for heterogeneous co-integration of new function with CMOS, *Journal on Emerging and Selected Topics in Circuits and Systems*, 2, 714–722, 2012.

47. P. Coudrain, Setting up 3D sequential integration for back-illuminated CMOS image sensors with highly miniaturized pixels with low temperature fully-depleted SOI transistors, *IEEE IEDM*, 2008.

48. M. Vinet et al., Monolithic 3D Integration: A powerful alternative to classical 2D scaling, *IEEE S3S*, 2014.

49. L. Brunet et al., Direct bonding: A key enabler to CoolCube™ (sequential) integration, *IEEE ECS*, 2014.

50. R. Choi et al., Bonding based channel transfer and low temperature process for monolithic 3D integration platform development, *IEEE S3S*, 2016.

51. L. Pasini et al., High performance low temperature activated devices and optimization guidelines for 3D VLSI integration of FD, Trigate, FinFET on insulator, *IEEE VLSI*, 2015.

52. C. Fenouillet-Beranger et al., Recent advances in low temperature process in view of 3D VLSI integration, *IEEE S3S*, 2016.

53. D. Benoit et al., Interest of SiCO low k = 4.5 spacer deposited at low temperature (400°C) in the perspective of 3D VLSI integration, *IEEE IEDM*, 2015.

54. C.-M.V. Lu et al., Dense N over CMOS 6T SRAM cells using 3D sequential integration, *IEEE VLSI-TSA*, 2017.

55. C.-M.V. Lu et al., Key process steps for high performance and reliable 3D sequential integration, *IEEE VLSI*, 2017.

56. F. Deprat et al., Technological enhancers effect on NiCo silicide stability for 3D sequential integration, *EMRS Spring*, 2016.

57. F. Deprat et al., NiCo 10%: A promising silicide alternative to NiPt 15% for thermal stability improvement in 3DVLSI integration, *AMC*, 2015.

58. M. Vinet et al., Opportunities brought by sequential 3D CoolCube™ integration, *IEEE ESSDERC*, 2016.

59. C. Fenouillet-Beranger et al., W and Copper interconnection stability for 3D VLSI CoolCube intergration, *SSDM*, 2015.

60. F. Deprat et al., Dielectrics stability for intermediate BEOL in 3D sequential, *MAM*, 2016.

61. A. Jain, R.E. Jones, R. Chatterjee, and S. Pozder, Analytical and numerical modeling of the thermal performance of three-dimensional integrated circuits, *IEEE Transactions on Components and Packaging Technologies*, 33(1), 56–63, 2010.

62. H. Xu, V.F. Pavlidis, and G. De Micheli, Analytical heat transfer model for thermal through-silicon vias, in *Design, Automation & Test in Europe (DATE)*, 2011, pp. 1–6.

63. C. Santos, P. Vivet, J.P. Colonna, P. Coudrain, and R. Reis, Thermal performance of 3D ICs: Analysis and alternatives, in *IEEE International 3D Systems Integration Conference (3DIC)*, 2014, pp. 1–7.

64. P. Coudrain, P. Momar Souare, J.P. Colonna, H. Ben-Jamaa, P. Vivet, R. Prieto, V. Fiori et al., Experimental insights into thermal dissipation in TSV-based 3D integrated circuits, *IEEE Design & Test*, 33, 21–36, 2015.

65. C. Santos, P.M. Souare, P. Coudrain, J.P. Colonna, F. de Crecy, P. Vivet, A. Borbely et al., Using TSVs for thermal mitigation in 3D circuits: Wish and truth, *3DIC*, December 2014, Cork, Ireland.

66. P.M. Souare et al., A comprehensive platform for thermal studies in TSV-based 3D integrated, *IEEE IEDM*, 2014.

67. S.K. Samal et al., Fast and accurate thermal modeling and optimization for monolithic 3D ICs, in *Design Automation Conference (DAC)*, 2014, pp. 1–6.

68. C. Santos, P. Vivet, S. Thuries, O. Billoint, J.-P. Colonna, P. Coudrain, and L. Wang, Thermal performance of CoolCube™ monolithic and TSV-based 3D integration processes, *3DIC'2016 Conference*, December 2016, San Francisco, CA.
69. P. Vivet, et al., A 4 × 4 × 2 homogeneous scalable 3D network-on-chip circuit with 326 MFlit/s 0.66 pJ/bit robust and fault tolerant asynchronous 3D links, *IEEE Journal of Solid-State Circuits*, 2016, 1–17.
70. P. Emma et al., 3D stacking of high-performance processors, in *International Symposium on High Performance Computer Architecture (HPCA)*, 2014, pp. 500–511.
71. Physical Verification with Calibre™. https://www.mentor.com/products/ic_nanometer_design/verification-signoff/physical-verification/.

8

Novel Platforms and Applications Using Three-Dimensional and Heterogeneous Integration Technologies

Kuan-Neng Chen, Ting-Yang Yu, Yu-Chen Hu, and Cheng-Hsien Lu

CONTENTS

8.1 Introduction

With the growing demands of high computing and Internet of Things (IoT) applications, high I/O counts with fast signal transmission speed and variety integration ability become equally significant in current semiconductor development. Platforms based on 3D-advanced packaging and integration can provide such solutions to high-speed, low-power, small form factors, and heterogeneous integration requirements. Advanced applications that have been successfully manufactured such as graphics processor units (GPU) and mobile processers adopt key technologies of 3D integration and advanced packaging such as fan-out scheme, whereas advanced complementary metal–oxide–semiconductor (CMOS) image sensors use through-silicon via (TSV) and fine-pitch Cu direct bonding. Other than these applications, 3D integration and heterogeneous technologies create many opportunities and a whole new world for advanced systems, which were difficult to be fulfilled in the past. This chapter describes examples of novel platform and application demonstration, including optical components, pressure sensors, and bioneural applications.

8.2 Stacked Terahertz Optical Component

8.2.1 THz Wave Applications

The *THz gap*, which ranges between ~0.1 and ~10 THz, is a gap in the electromagnetic spectrum for generating and detecting the radiation that does not live because of rare light source. Luckily, the bottleneck of producing THz has been gotten through free-electron laser, nonlinear effect of electro-optic (EO) crystal, quantum cascade, photoconductive switch, and so forth lately [1,2]. THz radiations are nondestructive and intrinsically safe compared to X-ray because of low photon energy. Furthermore, it is highly sensitive with metals, some chemical bond and function groups, and therefore THz wave is broadly utilized for material characterizations, security, and bio-images. Apart from these, THz astronomical observations are also a vital application for water and dust detection from space. Also, THz is the carrier frequency of next generation for communications. Due to THz source breakthrough, the quantities of THz applications have embraced a boom. However, researchers have not systematically explored the relative optical components, such as polarizer and filter. In this endeavor, the prior art of THz polarizer fabrication by 3D-IC technologies would be discussed and compared by other methods.

8.2.2 Issues of Common THz Polarizers

Structures of optical components are usually with scale in wavelength or subwavelength. Therefore, THz wire-grid polarizers are usually with pitch of microns. The commercial wire-grid polarizer stands in air or on the thin film, and also known as free-standing or thin-film wire-grid polarizer. The transverse magnetic (TM) mode of THz wave would infiltrate the polarizer without power loss because of free-standing structure. However, the free-standing wire-grid polarizers are very fragile without substrate supporting. The mechanical strength of polarizer would be improved by fabricating on substrate but enhances reflective loss. To align, the polarizer on substrate at Brewster's angle would increase the transmission power to 100%. However, specific angle alignment is difficult to modulate in systems. Antireflection (AR) coating is another option to reduce reflection loss. Single-layer AR coating with thickness of quarter wavelength and suitable refractive index would enhance the transmittance to near 100%. Figure 8.1 demonstrates the reasoning stream and composed structure of new sort polarizer. The AR coating

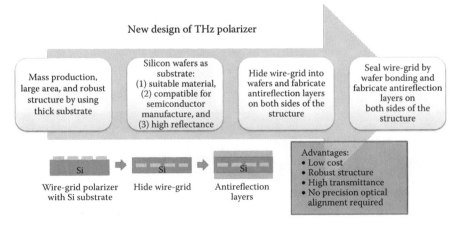

FIGURE 8.1
The designed diagram and advantages of new type THz polarizer.

could be fabricated on both substrate surface, if the wire-grid was sealed in to the substrate. The power transmittance would be theoretically 100%. This section would demonstrate the seal of wire-grid polarizer by wafer bonding and AR-coating process.

8.2.3 Fundamentals and Fabrication Methods

8.2.3.1 Structure Design of THz Polarizer

Silicon is a most suitable substrate for wire-grid polarizer, which has advantages of robustness, low cost, low dispersion, and low absorption loss in THz [3]. However, in THz region, silicon substrate usually accompanies high reflectance because of the high refractive index. In order to avoid high reflectance due to Si substrate, AR coating must be designed to have precise thickness (t_{AR}) of a quarter of wavelength and refractive index (n_{AR}) of square root of Si refractive index for a single frequency. The purpose of AR coating is to make the first reflection beam from surface and second reflection beam from interface between AR coating and Si destructive interference, as shown in Figure 8.2. The relations between t_{AR} and n_{AR} are shown in the following equations:

$$t_{AR} = \frac{\lambda_0}{4n_{AR}} \tag{8.1}$$

$$n_{AR} = \sqrt{n_{si}} = \sqrt{3.4} = 1.84 \tag{8.2}$$

where λ_0 is the central wavelength in free space. According to Equations 8.1 and 8.2, the AR-coating material must be in the range of microns thickness and refractive index of 1.84. But in THz region, it is really tough to find such a material with exact refractive index and low absorption loss. Moreover, microns thickness coating layer would induce large stress between AR

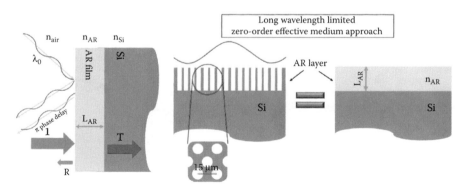

FIGURE 8.2
AR coating with quarter wavelength thickness leads to destructively interfere, and the index could be controlled by whole array pitch under zero-order effective medium approach.

coating and substrate, and also get peeled. Therefore, Si substrate is etched by deep reactive-ion etching (DRIE) with cylinder holes and the index of AR layer is tuned to specific index and also specific thickness according to the effective medium theory. The effective index of two mixing materials is equal to Equation 8.3:

$$n_{AR} \cong f \times 3.4 + (1-f) \times 1 = \sqrt{3.4} \tag{8.3}$$

where f is the filling factor of Si and air and is 0.35 to achieve the effective index. As the pitch of etching holes is less than $\lambda_0/10$, the AR layer seems like a homogeneous layer with effective index, as shown in Figure 8.2.

8.2.3.2 Fabrication Methods and Low-Temperature Eutectic Liquid Bonding

The process flow of THz wire-grid polarizer fabrication is shown in Figure 8.3a. Double-side polished wafers are ready for etching and deposition of AR layer soon after standard Radio Corporation of America (RCA) clean. After that, AR layer is patterned through lithography and etched via DRIE with the optimized parameters. The diameter and space between hole array are etched to 12.6 and 2.4 μm with hexagonal arrangement according to

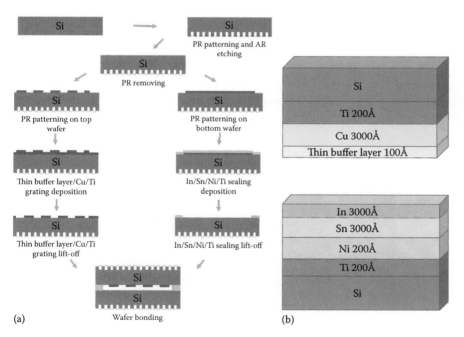

FIGURE 8.3

(a) The process flow of stacked THz polarizer fabrication (From Chi, N.-C. et al., *IEEE ECTC*, 1793–1798, 2017.) and (b) the bonding structure for Cu wire-grid sealing. (From Chi, N.-C. et al., *SPIE Optics + Optoelectronics*, 102420Z–102420Z-6, 2017.)

the designed filling factor and AR thickness. The Cu wire-grid gratings and low-temperature In/Sn solder-sealed rings are deposited with relative thickness, which are patterned on the other side of the respective Si wafers after fabricating the AR layer, as shown in Figure 8.3b. Two wafers are bonded face to face at 150°C for 30 minutes. The bonding mechanism uses low eutectic point (118°C) of In/Sn alloy. Therefore the polarizer could be bonded at low temperature [4]. Using ultrathin buffer layer could tentatively avoid Cu-In/Sn interdiffusion and ensure that the residue of In/Sn would melt even with submicron thickness [5]. The fabrication method of robust THz polarizer with AR layers using 3D-IC technologies is demonstrated. The Cu wire-grid is sealed into bonded wafers to prevent corrosion.

8.2.4 Performance of Staked THz Polarizer

8.2.4.1 Bonding and Etching Qualities

When the temperature rises to the eutectic point of In/Sn binary alloy, which is 118°C, In interdiffuses with Sn at the interface and then melts at the interface. The liquid layer of In/Sn improves the bonding yield because of large tolerance of surface roughness. However, In/Sn solder also interdiffuses with Cu to form intermetallic compound (IMC) immediately. The residue of In/Sn solder is not enough to form the liquid layer if the solder layer is too thin, which could cause bonding failure. A Ni buffer layer is capped on Cu layer as a barrier to delay the interdiffusion of the solder and Cu. To test the bonding yield of the bonding structure in this research, blanket wafers with the same metal bonding layers as mentioned earlier are fabricated as shown in Figure 8.3b and bonded with the same condition of the polarizers. Figure 8.4a shows good bonding yields of Cu-In/Sn bonding structure by scanning acoustic tomography (SAT) images of the blanket sample.

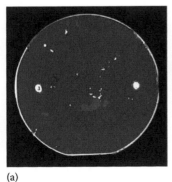

Size: 2 cm × 2 cm
12 polarizers for a bonded wafer

(a) (b)

FIGURE 8.4
(a) SAT image showing the excellent bonding quality and (b) the finished component of THz polarizers. (From Yu, T.-Y. et al., *SPIE Optical Engineering + Applications*, 95850L–95850L-7, 2015.)

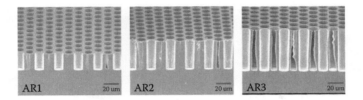

FIGURE 8.5
SEM images and parameters set up of AR layers. (From Yu, T.-Y. et al., *SPIE Optical Engineering + Applications*, 95850L–95850L-7, 2015.)

As discussed earlier, the depth and filling factor of the etching holes control the central frequency and the refractive index of the AR layer. Figure 8.5 demonstrates the designed parameters and SEM measurement results of the three different AR layers. Comparing the etching results with the designed parameters, all samples are close to the original design except the hole diameter of AR3, which is a little larger. The large holes make the n_{AR} smaller than the designed value because of smaller filling factor according to Equation 8.3. Modified etching process would improve the etching profile control.

8.2.4.2 High Transmittance THz Polarizer

THz time-domain spectroscopy (TDS) is used to measure the transmittance of stacked THz polarizers. The samples are placed behind a commercial free standing wire-grid polarizer that fixes polarization direction during the measurement. Figure 8.6 shows the measured method of the THz-TDS

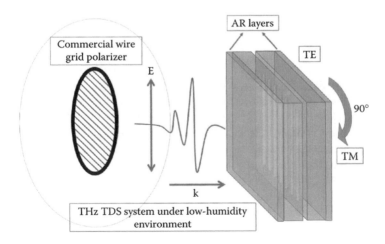

FIGURE 8.6
Stacked THz polarizer measurement by THz-TDs through commercial polarizer. (From Yu, T.-Y. et al., *SPIE Optical Engineering + Applications*, 95850L–95850L-7, 2015.)

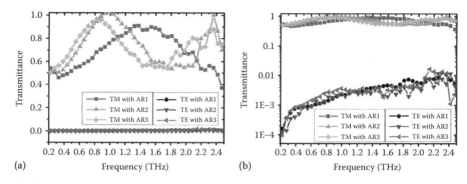

FIGURE 8.7
(a and b) The transmittance spectrum of TE and TM mode of stacked THz polarizers in linear and log scale. (From Yu, T.-Y. et al., *SPIE Optical Engineering + Applications*, 95850L–95850L-7, 2015.)

measurement. THz-TDS results demonstrate electric field variation in time domain. Fast Fourier transform (FFT) transfers TDS signal to the frequency domain with complex amplitude and phase information and then transfers to transmittance spectrum after a simple calculation. Figure 8.7a shows the transmittance spectra of transverse electric (TE) and TM mode of stacked polarizers. Over 90% of TM-mode transmittance and minimum transmittance about 0.01% of TE mode are demonstrated with each polarizer sample. Especially, the AR2 sample reaches almost 100% TM transmittance at the designed central frequency. However, the enlarged holes of AR3 sample shift the central frequency to higher frequency as a result of smaller effective index. Nevertheless, all three samples show high extinction ratio between 20 and 40 dB from 0.2 to 2.2 THz as shown in Figure 8.7b.

8.2.4.3 Broadband THz Polarizer

For the broadband THz polarizer, two wafers with different AR layers are bonded and the Cu wire-grid is sealed by bonded wafers. The transmittance spectrum is broadened due to combination of two different AR layers with different central frequencies. Figure 8.8 shows the transmittance spectrum of different stacked AR layers, and the transmittance is basically the product of transmittance of each AR layer. Therefore, the bandwidth of transmittance spectrum is extended more than 1 THz. Furthermore, the peak values of AR1×AR2 and AR1×AR3 polarizers have uniform transmittance about 85% and 70% with width more than 250 GHz and 1 THz, as shown in Figure 8.8b and d. This means when passing through the broadband polarizer the

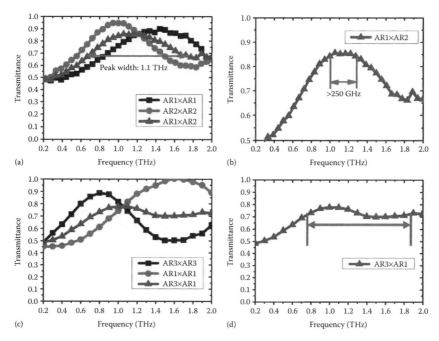

FIGURE 8.8
Transmittance spectra of (a) TE and (b) TM mode of stacked THz with AR1×AR2 layers, and (c) TE and (d) TM mode of THz with the AR1×AR3 layers. (From Chi, N.-C. et al., *SPIE Optics + Optoelectronics*, 102420Z–102420Z-6, 2017; Chi, N.-C. et al., *IEEE ECTC*, 1793–1798, 2017.)

broadband signal is distortion less and still has high-power transmission. Hence, the THz polarizer fabricated by bonding two different AR layers could have a wider usage rather than limited in central frequency owing to its narrow bandwidth.

8.2.5 Comparison between Common and Stacked THz Polarizers

Table 8.1 demonstrates the previous works of THz polarizers and characteristic of stacked THz polarizer [8]. Stacked polarizer has advantages with large area, low cost, high performance, and convenient for using. Most materials of THz polarizers listed in the table are not so easy to fabricate or easy to use. However, stacked polarizer is almost fabricated on Si and really potential for manufacturing because of mature semiconductor foundries. In conclusion, stacked THz polarizer provides a better option for THz optical system, which comparably requires robust structure but provides efficient performance.

TABLE 8.1

Overview of Some Designed and Commercial THz Polarizer

Group	Transmittance	Extinction Ratio (dB)	Measured Bandwidth (THz)	Structure
Hsieh et al.	0.60 ~ 0.85	20 ~ 40	0.2 ~ 1.0	Feussner polarizer based on a liquid crystal
Ren et al.	0.75 ~ 0.95	18 ~ 50	0.2 ~ 2.2	Carbon nanotube (CNT) layers
Kyoung et al.	0.45 ~ 0.60	27 ~ 38	0.1 ~ 2.0	Reel-wound CNT multilayers
Wojdyla and Gallot	0.90 ~ 0.99	37.8 (max)	N/A	Brewster's polarizer using a stack of silicon wafers
Yamada et al.	0.95	35 ~ 50	0.5 ~ 3.0	Al on thick Si substrate angled at Brewster angle
MacPherson et al.	0.91 ~ 0.98	20 ~ 37	0.2 ~ 2.0	Thin-film aluminum on SiO_2
Microtech	0.9 ~ 1.0	20 ~ 40	0 ~ 3	Free standing wire grid
Microtech	0.6 ~ 1.0	20 ~ 40	0 ~ 3	Wire-grid pattern on thin substrate
Tydex	0.9 ~ 0.95	~20	0 ~ 3	Wire-grid pattern on thin substrate
Stacked THz polarizers	0.5 ~ 1.0	20 ~ 40	0.2 ~ 2	Wire-grid sealed into Si wafer and AR processing on surface

Source: Yan, F. et al., *J. Infrared Millim. Terahertz Waves*, 34, 489–499, 2013.

8.3 Pressure-Sensing System

Heterogeneous integration is an important development direction for IoT cloud and smart life [9]. Furthermore, complementary metal–oxide–semiconductor–microelectromechanical systems (CMOS–MEMS) integration is critical in the future to sense the external environment. Advanced packaging technologies are developed to integrate different functions of the chips including sensor, logic, and memory. Among the packaging approaches, 3D/2.5D integration technologies have attracted attention because they are highly integrated, low weight, with high performance, and small form factor [10–11]. However, serious issues still exist in conventional sensor chip of 3D/2.5D integration. Due to the functionality of sensor chip, the surface morphology is quite different from circuit chip. The surface conditions of sensor chip usually have complex topography and brittle surface, even transparent material. Hence, the handling of sensor chip suffers throughput and alignment accuracy and becomes a bottleneck during assembly. Therefore, an alternative approach is needed to overcome the limitations.

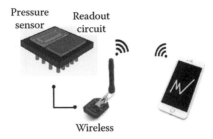

FIGURE 8.9
Concept of pressure-sensing system for monitoring external pressure variation. (From Hu, Y.-C. et al., *IEEE IEDM*, 9.2.1–9.2.4, 2016.)

Self-assembly technology has been proposed for multichip integration to improve the throughput [12–13]. This approach is also suitable for the hard-to-handle sensor chip. Furthermore, we develop a double-self-assembly technology to implement chip-level stacking with more complicated topography. A pressure sensing system is adopted to verify this technique as shown in Figure 8.9. A heat dissipation micropin-fin interposer with uneven backside profile is formed to carry a readout circuit chip and a complex topography sensor chip. Cu/In metal bonding is used for low resistance and high I/O density in this platform. The excellent reliability electrical tests prove that the innovative double-self-assembly approach can efficiently integrate CMOS–MEMS chips efficiently.

8.3.1 Pressure-Sensing Platform

The block diagram of the integrated pressure-sensing platform is mainly composed of four parts: pressure sensor, readout circuit, wireless module, and a receiver, which is shown in Figure 8.10. For small form factor and lower power consumption, the pressure sensor and readout circuit chip

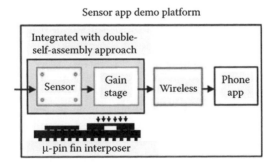

FIGURE 8.10
Pressure-sensing system consists of sensor chip, readout circuit chip, wireless module, and receiver. (From Hu, Y.-C. et al., *IEEE IEDM*, 9.2.1–9.2.4, 2016.)

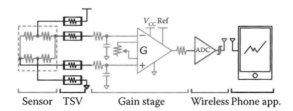

FIGURE 8.11

Block diagram of the proposed pressure-sensing system. The signal begins from the pressure sensor to the receiver. (From Hu, Y.-C. et al., *IEEE IEDM*, 9.2.1–9.2.4, 2016.)

are integrated on the heat dissipation micropin-fin interposer. Double-self-assembly approach is adopted to assemble the stacked chips. Figure 8.11 presents the simplified architecture and equivalent circuit of the proposed pressure sensing system. The integrated platform transmits voltage signal from the pressure sensor to readout circuit through TSV and redistribution layer (RDL) on the interposer under pressure variation. The readout circuit is mainly configured a gain stage, which amplifies weak signal to an analog-to-digital converter (ADC) circuit. The wireless chip receives and transmits information to a mobile device receiver. Users can be remotely informed the environmental pressure variation on the screen.

8.3.2 Pressure Sensor

A piezoresistive pressure sensor is adopted in the pressure-sensing system for sensing environment pressure difference. When the piezoresistive material is deformed due to a change in stress, the resistance value also changes with the characteristic of piezoresistive material. Therefore, we can measure the pressure through the circuit connection and a designed Wheatstone bridge. Figure 8.12 shows SEM image of the pressure sensor and a schematic of the integrated pressure system. A membrane on the pressure sensor is fabricated by silicon DRIE and ion implantation for sensing the pressure variation. However, the top side of the pressure sensor is fragile due to slender membrane structure, and the backside of the pressure is hard to handle as the surface is uneven with cavities.

8.3.3 Micropin-Fin Heat Sink Interposer

Interposer is adopted for carrying CMOS readout circuit chip and pressure sensor chip. The particular design, micropin-fin, is developed for heat dissipation due to heat accumulation, which is a critical issue in 3D IC integration.

Figure 8.13 shows the schematic and SEM image of the micropin-fin heat sink interposer. A 200 mm silicon wafer is processed to fabricate Cu TSV with the formation of 30 μm diameter and 250 μm pitch. On the front side of

FIGURE 8.12
Schematic of the integrated readout circuit chip, pressure sensor chip, and micropin-fin inter-poser. Cross-sectional and top view of the pressure sensor. (From Hu, Y.-C. et al., *IEEE IEDM*, 9.2.1–9.2.4, 2016.)

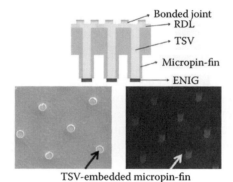

FIGURE 8.13
Schematic of the micropin-fin heat sink interposer and SEM image of the micropin-fin. (From Hu, Y.-C. et al., *IEEE IEDM*, 9.2.1–9.2.4, 2016.)

interposer, two RDLs are fabricated for connecting sensor chip and circuit chip. Cu/In-bonded pads are formed on the RDL. On the backside of inter-poser, electroless nickel immersion gold (ENIG) is chosen as bonded pads. DRIE process is adopted to etch deeply micropin-fins shape. The relation-ship between micropin depth and temperature reduction is simulated. TSV daisy chain electrical characteristics, thermal cycling test (TCT) with −55°C to 125°C for 750 loops, and highly accelerated stress test (HAST) at 130°C for 96 hours have been investigated [14].

8.3.4 Integration of Micropin-Fin Heat Sink Interposer and Chips with Double-Self-Assembly Approach

Double-self-assembly approach is developed and adopted to integrate CMOS readout circuit chip, pressure sensor chip, and micropin-fin heat sink interposer. Figure 8.14 illustrates the 3D/2.5D assembly process flow between the chips and interposer. On the carrier wafer, a 500 nm hydrophilic oxide is deposited, followed by patterned perfluorodecyltrichlorosilane (FDTS) hydrophobic film to define the self-assembly area of the interposer. Afterwards, precisely controlled small volume of liquid is dropped on the hydrophobic area and interposer is placed roughly on the liquid. The micropin-fin side of interposer is also covered with oxide layer for self-assembly expect ENIG-bonded pad. As the interposer is placed on the liquid, the liquid surface tension pulls and drags the interposer to the defined hydrophilic area immediately. Figure 8.14a demonstrates this procedure of the interposer first self-assembly temporary bonding on the carrier wafer. The interposer front side is also patterned FDTS hydrophobic film for the readout circuit and sensor chip locations with bonded pads. After the chips are placed on top of the liquid, the two chips are also self-aligned to the defined area with the same method and mechanism. Next, the continued thermo-compression bonding is adopted to accomplish the second self-assembly permanent bonding of the readout circuit chip and sensor chip on the micropin-fin interposer, as shown in Figure 8.14b. Finally, the 3D/2.5D pressure-sensing assembled chip is detached from the carrier wafer once the liquid gets evaporated.

In order to measure the alignment accuracy, a designed misalignment measurement mark with 0.5 µm accuracy is implemented to study the relationship with optimized liquid volume. The results of optimized liquid volume of four different heights of micropins suggest that filling up each 10 µm depth of micropin gap must require 0.25 µL of liquid. The micropin-fin interposer size depends on the liquid volume during first self-assembly

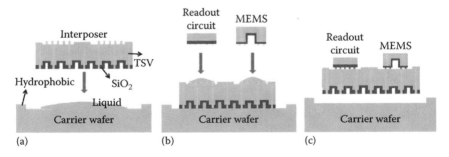

FIGURE 8.14
Process flow of double-self-assembly approach: (a) Interposer is temporarily attached on the carrier wafer for the first self-assembly, (b) readout circuit and sensor chips are permanent attached on the interposer for the second self-assembly, and (c) the integrated assembly chips are detached from the carrier wafer. (From Hu, Y.-C. et al., *IEEE IEDM*, 9.2.1–9.2.4, 2016.)

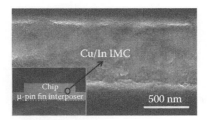

FIGURE 8.15
Cross-sectional SEM image of Cu/In-bonded joint with no void or seam. (From Hu, Y.-C. et al., *IEEE IEDM*, 9.2.1–9.2.4, 2016.)

temporary bonding on the carrier wafer. The second self-assembly experiments are performed with five different chip sizes to assess the corresponding optimized liquid volume [14].

Figure 8.15 illustrates the cross-sectional SEM image of Cu/In between upper die and lower interposer after double-self-assembly. From the image, there are no crack and void at the bonding interface. In addition, SEM, SAT, and EDX are used to investigate the bonding quality. Based on EDX results, Cu-In is completely transferred into IMC. Shear test is adopted for mechanical strength analysis. Cu-In bonding strength is over 20 MPa [14]. The material analysis and mechanical test prove the robust structure with the double-self-assembly approach.

The electrical characteristics and reliability tests are evaluated using a Kelvin structure consisting of Cu/In thin-film metal pad-bonded joint to prove the feasibility of double-self-assembly method. This design is used to measure the specific contact resistance of bonded joint. The obtained resistance value from Kelvin structure is around $10^{-9}\ \Omega\cdot cm^2$. HAST and TCT are carried out for 96 hours with 85% RH and 750 loops under −55°C to 125°C, respectively. Results of stable specific contact resistance before and after HAST imply good moisture resistance capability with double-self-assembly approach. TCT also indicates good electrical property under significant changes in temperature. The stable reliability results of HAST and TCT of the integration between chips and interposer demonstrate the stability and feasibility of the double-self-assembly approach. Figure 8.16 shows the proposed 2.5D/3D integrated pressure-sensing system, which contains CMOS readout circuit chip, pressure sensor chip, and micropin-fin heat sink interposer with double-self-assembly approach.

8.3.5 Achievements and Outlook

A novel 3D integration technology based on liquid-surface tension mechanism to solve difficult-to-handling chip issue of double-self-assembly approach is introduced. Using this advanced technique, we demonstrate a pressure-sensing system with readout circuit chip, complex topography

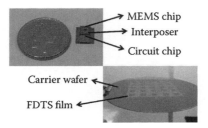

FIGURE 8.16
Photograph of the proposed integrated pressure-sensing system with double-self-assembly approach. (From Hu, Y.-C. et al., *IEEE IEDM*, 9.2.1–9.2.4, 2016.)

sensor chip, and micropin-fin interposer. Chips with Cu-In bonded pads are successfully self-assembled on the interposer. Moreover, interposer is also temporary attached on the carrier wafer and well debonded with first self-assembly approach. Various sizes of interposer and chip are investigated with the relationship of optimized liquid volume. The resulting electrical characteristic shows high-accuracy alignment and stable reliability, which implies the feasibility of double-self-assembly. Heterogeneous integration can be further extended to various types by which difficult-handling chip can be processed based on this novel technology.

8.4 Neurosensing Systems

In 1942, people first recorded the feeble brain signal from the scalp. Since then, people realized the importance of relationship between electrical signal and the diseases using electrocorticography (ECoG) or electroencephalography (EEG) to record the brain signal. The former president of the United States of America, Barack Obama, has proposed the national research strategy "Brain Research through Advancing Innovative Neuro technologies" in 2013. Thus, the national research institutes throughout the world put lots of effort to break through the clinical trial therapy. In order to complete understanding the relationship between the brain signal and diseases, high spatial brain signal extraction accuracy, low-signal transmitting loss, and high-resolution neural sensing biosensor are very important. In recent years, different types of neural sensing biosensors have been proposed, including electrical signal analysis circuit chips, cables or wire bonding for signal transmission, and passive detection devices. However, there are some issues in the traditional neural sensing biosensor [15]. First, wire-bonding connection and cables may cause weak brain signal degradation and noise [16]. Second, spatial resolution limits by the electrode channel numbers and its process [17–18]. Third, the yield is low because of complicated process for electrode and circuit chip in one wafer.

Three-dimensional integration technology is a potential packaging scheme to overcome the above-mentioned issues to realize the novel neural sensing biosensors. TSV and metal thin-film bonding techniques can substitute the long cables and wire bonding to reduce signal degradation. Utilizing standard semiconductor fabrication for silicon interposers and electrode probes can achieve higher channel numbers and lower cost. Moreover, three-dimensional heterogeneous integrations with bonding process can combine optimized functional circuit chips, electrodes, and substrates, which can increase yield, and the process is much more flexible.

In this section, three types of neural sensing biosensors, 3D-SiP neural sensing biosensor, 2.5D-silicon interposer neural sensing biosensor, and 2.5D-flexible interposer neural sensing biosensor, are proposed for different functional aspects.

8.4.1 Fabrication, Scheme, and Reliability of Neural Sensing Biosensor

8.4.1.1 Three-Dimensional System-in-Packaging Neural Sensing Biosensor

The scheme of 3D system-in-packaging (SiP) neural sensing biosensor contains TSV-embedded electrode array, silicon interposer, and two stacked circuit chips. The chip-level integration scheme is capable of integrating different substrates, chip sizes, and technology nodes by using suitable bonding materials. Furthermore, it can be applied to wafer-level integration scheme.

8.4.1.2 Chip-Level Heterogeneous Integration Scheme

In this scheme, two stacked circuit chips are fabricated by Taiwan semiconductor manufacturing company (TSMC) 40 nm and 0.18 μm CMOS node to reveal the feasibility of chip-level Cu/Sn joint to ENIG bonding. As utilized bump height of the joint must be less than 10 μm, the thicknesses of Cu and Sn are 4 and 5 μm, respectively. The thickness of ENIG is 3.25 μm, which contains 3 μm Ni and 0.25 μm Au.

There are 2035 μ-joints for the following face-to-face bonding process. The two chips are bonded together at 270°C for 10 seconds. Figure 8.17 shows the cross-sectional SEM image of chip level heterogeneous bonding without having any voids and defects.

8.4.1.3 Electroplating Solution Improvement

Along with the fine pitch bump, the reduction of bump height becomes more essential. Figure 8.18 illustrates the material phase transformation during the bonding process. The concave–convex phenomenon caused by the Cu/Sn bump formation and the passivation layer has been noticed in previous studies. However, the nonuniform surface causes degradation of the bonding quality in sub 10-μm-bumps. This issue can be improved by chemical–mechanical

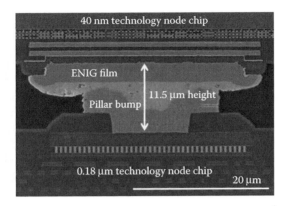

FIGURE 8.17
SEM cross-section image of chip-level heterogeneous integration scheme. (From Hu, Y.-C. et al., *IEEE Trans. Electron Devices*, 62, 4148–4153, 2015.)

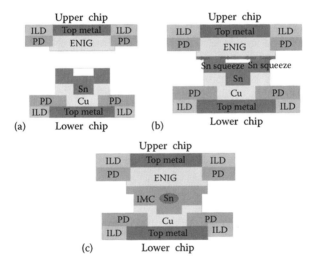

FIGURE 8.18
The schematic of material-phase transformation during bonding process. (a) ENIG film and concave-shaped bump formation before bonding, (b) During bonding, Sn squeezes outside and inside the concave-shaped bump, (c) Pure Sn is surrounded by the middle of IMC. (From Hu, Y.-C. et al. *IEEE Trans. Electron Devices*, 62, 4148–4153, 2015.)

planarization (CMP) process, which is very expensive and dirty. In this study, by adjusting the plating current and changing plating solution, the bonding quality can be improved efficiently. The adopted methodology is comparably lesser in cost, highly efficient, and also compatible to semiconductor processes. The plating results under different plating current and solution shows that the height difference between the bump center and edge is less than 1 μm [19]. During the plating process, the chloride in the improved solution links with the Cu$^+$ ion to form the suppression layer on the cathode. Thus, the Cu^{+2} on the surface of the cathode will not be inhibited.

8.4.1.4 Electrical Reliability Tests

The electrical performance and reliability of the bonded structure are tested by Kelvin structure and daisy chain. One bonded joint resistance is ~0.22Ω by Kelvin structure measurement [20]. A 30-daisy-chain in series of ENIG to Cu/Sn-bonded joints has a resistance of 24.6Ω [20]. The reliability tests used here are TCT and unbiased HAST, which test the thermal reliability of the bonded joint and the moisture resistance. Results of a 1000-loops TCT test, which is under −55°C to 125°C with 15°C/min ramping rate show that the resistance of the bonded joints gets better because of the thermal treatment during the thermal cycling that enhances the diffusion to eliminate the bonding interface and may also form low-resistance IMC [20]. The 96-hours unbiased HAST under 85% RH and 130°C results illustrate good humidity resistance. Thus, the heterogeneous-bonded structure shows good bonding quality and reliability [20].

8.4.2 2.5D-Silicon Interposer Neural Sensing Biosensor

In order to realize the high spatial and resolution neural sensing biosensor, the channel numbers should be increased and cost must be scaled down. The 2.5D-silicon interposer neural sensing biosensor can achieve all the demands, as shown in Figure 8.19. This scheme illustrates the capability of heterogeneous integration, different functional chips, high channel transition, and semiconductor fabrication compatibility.

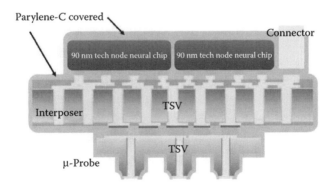

FIGURE 8.19
Schematic of 2.5D-silicon interposer neural sensing biosensor. (From Hu, Y.-C. et al., *IEEE Trans. Electron Devices*, 64, 1666–1673, 2017.)

8.4.2.1 Fabrication of Silicon Interposer and Through-Silicon Via-Embedded µ-Probe

Figure 8.20 shows the process flow of the TSV of silicon interposer. The diameter and depth of TSV are 30 and 130 µm, respectively. First, utilizing Bosch process to etch the silicon via and deposit 1 µm thickness oxide liner to prevent the diffusion of atom and leakage current. Second, depositing TiN adhesive layer and Cu seed layer for the following Cu electroplating process. After the electroplating process, CMP process will be conducted to remove the superfluous Cu film. The wafer is grinded to 130 µm, and then the RDL and ENIG bonding pad are fabricated. Finally, two-layer RDL and 50-µm µ-bump on another side for the functional circuit chips bonding joint are accomplished.

TSV-embedded electrode µ-probe fabrication process flow is shown in Figure 8.21. After TSV fabrication, photoresist defines the pitch and the size of probe. The isotropic etching uses SF$_6$ to shape the probe opening. Then the nonisotropic inductively coupled plasma (ICP) etching is employed to define the pitch depth. Pitch, diameter, and depth of the µ-probe are 140, 70, and 80 µm, respectively. Finally, Ti and Pt thin film layers are deposited on the probe to carry out the biocompatible electrode.

8.4.2.2 Electrical Reliability Test

The Cu TSVs and the bonded joints are verified under TCTs and unbias HAST to check the electrical reliabilities, including the 400 Cu TSVs in series

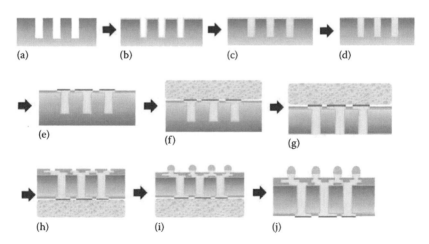

FIGURE 8.20
Process flow of silicon interposer. (a) TSV etching, (b) oxide linear and seedlayer deposit, (c) TSV plating, (d) Cu CMP, (e) RDL and ENIG pad fabrication, (f) temporary bonding, (g) wafer thinning, (h) 2 RDLs fabrication, (i) microbump plating, and (j) handle wafer debond. (From Hu, Y.-C. et al., *IEEE Trans. Electron Devices*, 64, 1666–1673, 2017.)

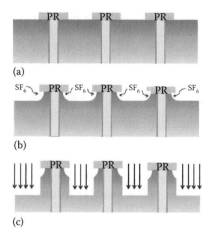

(a)

(b)

(c)

FIGURE 8.21
Process flow of TSV-embedded μ-probe. (a) PR is patterned as followed etching hard, (b) isotropic etching for concave shape around probe opening, and (c) Bosch process is adopted for probe formation. (From Hu, Y.-C. et al., *IEEE Trans. Electron Devices*, 64, 1666–1673, 2017.)

resistance and bonded joints resistance before and after the 1000 thermal cycling loops, and the Cu TSVs in series and bonded joints resistance before and after the unbias HAST under 130°C, 85% RH for 96 hours. All the reliability test results show good reliability of the 2.5D neural sensing biosensor scheme [18].

8.4.3 2.5D-Flexible Interposer Neural Sensing Biosensor

In this section, the polyimide, with biocompatible and bendable properties, better for animal experiments, is used for interposer to replace the silicon interposer. The polyimide is flexible along the surface of the cortex or other tissues. Furthermore, the polyimide substrate is used widely in semiconductor fabrication. Figure 8.22 shows the integration schematic of 2.5D-flexible interposer neural sensing biosensor, which contains μ-probe, PI interposer, and circuit chips.

8.4.3.1 Fabrication of Flexible Interposer

The polyimide flexible interposer substrate, with a thickness of 75 μm, is bendable and has high chemical resistance and high mechanical strength. 30-μm-diameter TSVs are Cu electrochemical deposited using a bottom-up method. Then 50-μm-diameter and 1-μm-height In μ-bump arrays are fabricated to carry the circuit chips.

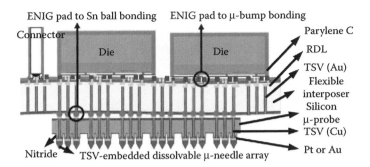

FIGURE 8.22
Schematic of 2.5D-flexible interposer neural sensing biosensor. (From Huang, Y.-C. et al., *Symposium on VLSI Technology*, pp. 218–219, 2016.)

8.4.3.2 Novel Flexible Bonding Approach

In order to realize the flexible neural sensing biosensor, the key part is silicon substrate to flexible substrate-bonding approach, which affects the electrical properties and the sensor reliabilities. Traditional bonding approach uses anisotropic conductive film (ACF) or nonconductive paste (NCP) in flexible substrate bonding. However, the number of the conductive particles must be well controlled for the good electrical connection without any leakage. Furthermore, the fine pitch is also limited by the size of conductive particles. Therefore, thin-film metal bonding approach is proposed to improve the bonding quality and to achieve ultrafine pitch.

The 300 nm Cu thin film and 10 nm Ti thin film are deposited on 75 μm polyimide flexible substrate, where Ti layer is the adhesive layer between the Cu film and polyimide substrate. Next, 20 nm Ni film as the buffer layer is deposited to prevent Cu oxidation before the bonding process. In order to achieve low-temperature bonding, 300 nm Sn film is used for the bonding material. On the other PI substrate, 300 nm In film is deposited. The two PI substrates are bonded together under a force of 100 N and at 170°C for 30 minutes. The cross-sectional SEM image of In/Sn-Cu-bonded joint reveals a good bonding quality [21].

8.4.3.3 Electrical Reliability Test of Novel Thin Film-Bonding Approaches

In this part, the electrical properties and reliabilities of two novel thin film metal-bonding approach are tested. The specific contact resistances of In/Sn-Cu thin film-bonding approach are measured under flat and bending radius of 30 mm by Kelvin structure, respectively. The specific contact resistance is about 2.5×10^{-7} $\Omega \cdot cm^2$ in both cases. The reliability performance of In/Sn-Cu thin film-bonding approaches under 500 thermal cycling loop of

TCT tests and the un-bias HAST under 130°C, 85% RH for 96 hours, respectively, illustrate the good bonded results, and the specific contact resistance is better than the conventional bonding approaches [20]. Thus, this bonding approach reveals the potential to improve the electrical properties of future flexible packaging.

8.4.4 Demonstration

The overall schemes, fabrication procedures, electrical, and reliability measurements of three types of the neural sensing biosensor schemes are discussed. Demonstrations of a 2.5D-silicon interposer neural sensing biosensor and a 2.5D-flexible interposer neural sensing biosensor are shown in Figures 8.23 and 8.24, respectively. Key technologies of 3D IC are used in these biosensors, such as heterogeneous integration, through silicon/flexible via, wafer level thinning, and chip-level bonding. Circuit chips, interposer, and TSV-embedded μ-probe are integrated within a small form factor. The reliabilities of the bonded joints and TSVs are also investigated. These results prove the potential of 3D IC technologies used in the biosensor scheme and other products in the near future.

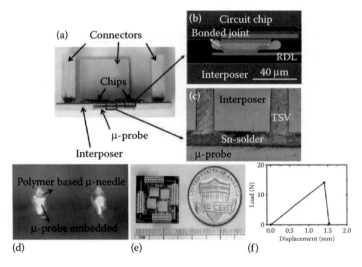

FIGURE 8.23
Photos of 2.5D-silicon interposer neural sensing biosensor. (a) Side view, (b) Cross-sectional view of circuit chip-interposer bonded joint, (c) Cross-sectional view of shunt-connected interposer-probe bonded joint, (d) μ-needle SEM image, (e) Photo of neural sensing microsystem, (f) Shear test diagram of the 2.5-D heterogeneous integrated neural sensing microsystem. (From Hu, Y.-C. et al., *IEEE Trans. Electron Devices*, 64, 1666–1673, 2017.)

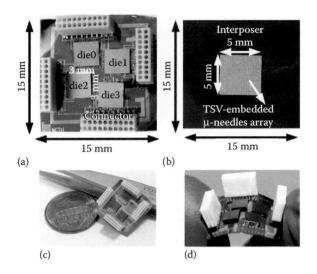

FIGURE 8.24
Photos of 2.5D-flexible interposer neural sensing biosensor. (a) Topview, (b) bottomview, (c) 256-ch microsystem, and (d) bending. (From Huang, Y.-C. et al., *Symposium on VLSI Technology*, pp. 218–219, 2016.)

References

1. R. A. Lewis, A review of terahertz sources, *Journal of Physics D: Applied Physics*, 47, 374001, 2014.
2. X. Yin, B. W.-H. Ng, and D. Abbott, Terahertz sources and detectors, in *Terahertz Imaging for Biomedical Applications: Pattern Recognition and Tomographic Reconstruction*, New York: Springer, 2012, pp. 9–26.
3. T.-Y. Yu, H.-C. Tsai, S.-Y. Wang, C.-W. Luo, and K.-N. Chen, High transmittance silicon terahertz polarizer using wafer bonding technology, in *SPIE Optical Engineering + Applications*, 2015, pp. 95850L–95850L-7.
4. H.-W. Liang, T.-Y. Yu, Y.-J. Chang, and K.-N. Chen, Asymmetric low temperature bonding structure using ultra-thin buffer layer technique for 3D integration, in *2016 IEEE 23rd International Symposium on the Physical and Failure Analysis of Integrated Circuits (IPFA)*, 2016, pp. 312–315.
5. Y.-J. Chang, Y.-S. Hsieh, and K.-N. Chen, Submicron Cu/Sn bonding technology with transient Ni diffusion buffer layer for 3DIC application, *IEEE Electron Device Letters*, 35, 1118–1120, 2014.
6. N.-C. Chi, T.-Y. Yu, H.-C. Tsai, S.-Y. Wang, C.-W. Luo, and K.-N. Chen, High transmittance and broaden bandwidth through the morphology of anti-relfective layers on THz polarizer with Si substrate, in *SPIE Optics + Optoelectronics*, 2017, pp. 102420Z–102420Z-6.
7. N.-C. Chi, T.-Y. Yu, H.-C. Tsai, S.-Y. Wang, C.-W. Luo, Y.-T. Yang, K.-N. Chen, High transmittance broadband THz polarizer using 3D-IC technologies, in *IEEE Electronic Components and Technology Conference (ECTC)*, 2017, 1793–1798.

8. F. Yan, C. Yu, H. Park, E. P. J. Parrott, and E. Pickwell-MacPherson, Advances in polarizer technology for terahertz frequency applications, *Journal of Infrared, Millimeter, and Terahertz Waves*, 34, 489–499, 2013.

9. W. Dehaene and A. S. Verhulst, New devices for internet of things: A circuit level perspective, in *2015 IEEE International Electron Devices Meeting (IEDM)*, 2015, pp. 25.5.1–25.5.4.

10. Y. Takemoto, K. Kobayashi, M. Tsukimura, N. Takazawa, H. Kato, S. Suzuki et al., Multi-storied photodiode CMOS image sensor for multiband imaging with 3D technology, in *2015 IEEE International Electron Devices Meeting (IEDM)*, 2015, pp. 30.1.1–30.1.4.

11. M.-H. Li, C.-Y. Chen, and S.-S. Li, A reliable CMOS-MEMS platform for titanium nitride composite (TiN-C) resonant transducers with enhanced electrostatic transduction and frequency stability, in *2015 IEEE International Electron Devices Meeting (IEDM)*, 2015, pp. 18.4.1–18.4.4.

12. T. Fukushima, T. Konno, K. Kiyoyama, M. Murugesan, K. Sato, W.-C. Jeong et al., New heterogeneous multi-chip module integration technology using self-assembly method, in *2008 IEEE International Electron Devices Meeting (IEDM)*, 2008, pp. 1–4.

13. T. Fukushima, E. Iwata, Y. Ohara, A. Noriki, K. Inamura, K.-W. Lee et al., Three-dimensional integration technology based on reconfigured wafer-to-wafer and multichip-to-wafer stacking using self-assembly method, in *2009 IEEE International Electron Devices Meeting (IEDM)*, 2009, pp. 1–4.

14. Y.-C. Hu, C.-P. Lin, H.-C. Chang, Y.-T. Yang, C.-S. Chen, and K.-N. Chen, An advanced 3D/2.5D integration packaging approach using double-self-assembly method with complex topography, and micropin-fin heat sink interposer for pressure sensing system, in *2016 IEEE International Electron Devices Meeting (IEDM)*, 2016, pp. 9.2.1–9.2.4.

15. Y.-C. Hu, Y.-C. Huang, P.-T. Huang, S.-L. Wu, H.-C. Chang, Y.-T. Yang et al., An advanced 2.5-D heterogeneous integration packaging for high-density neural sensing microsystem, *IEEE Transactions on Electron Devices*, 64, 1666–1673, 2017.

16. N. K. Logothetis, J. Pauls, M. Augath, T. Trinath, and A. Oeltermann, Neurophysiological investigation of the basis of the fMRI signal, *Nature*, 412, 150–157, 2001.

17. L.-C. Chou, S.-W. Lee, P.-T. Huang, C.-W. Chang, C.-H. Chiang, S.-L. Wu et al., A TSV-based bio-signal package with μ-probe array, *IEEE Electron Device Letters*, 35, 256–258, 2014.

18. C.-W. Chang, P.-T. Huang, L.-C. Chou, S.-L. Wu, S.-W. Lee, C.-T. Chuang et al., Through-silicon-via-based double-side integrated microsystem for neural sensing applications, in *IEEE International Solid-State Circuits Conference Digest of Technical Papers (ISSCC)*, 2013, pp. 102–103.

19. Y.-C. Hu, C.-P. Lin, Y.-S. Hsieh, N.-S. Chang, A. J. Gallegos, T. Souza et al., 3D heterogeneous integration structure based on 40 nm- and 0.18 μm- technology nodes, in *56th Electronic Components and Technology Conference (ECTC)*, 2015, pp. 1646–1651.

20. Y.-C. Hu and K.-N. Chen, A novel bonding approach and its electrical performance for flexible substrate integration, *IEEE Journal of the Electron Devices Society*, 4, 185–188, 2016.

21. Y.-C. Hu, C.-P. Lin, Y.-J. Chang, N.-S. Chang, M.-H. Sheu, C.-S. Chen, and K.-N. Chen, A novel flexible 3-D heterogeneous integration scheme using electroless plating on chips with advanced technology node, *IEEE Transactions on Electron Devices*, 62, 4148–4153, 2015.
22. Y.-C. Huang, Y.-C. Hu, P.-T. Huang, S.-L. Wu, Y.-H. You, J.-M. Chen et al., Integration of neural sensing microsystem with TSV-embedded dissolvable μ-needles array, biocompatible flexible interposer, and neural recording circuits, in *Symposium on VLSI Technology*, 2016, pp. 218–219.

Index

Note: Page numbers followed by f and t refer to figures and tables respectively.